Getting Started with 3D Carving

Five Step-by-Step Projects to Launch You on Your Maker Journey

Zach Kaplan

Getting Started with 3D Carving

by Zach Kaplan

Copyright © 2017 Zach Kaplan, Jeff Solin, Bob Clagett, David Picciuto, Travis K. Lucia, Casey Shea, Jimmy DiResta, Zachary Darmian-Harris, and Steve Carmichael. All rights reserved.

Printed in Canada.

Published by Maker Media, Inc., 1700 Montgomery Street, Suite 240, San Francisco, CA 94111.

Maker Media books may be purchased for educational, business, or sales promotional use. Online editions are also available for most titles (*http://oreilly.com/safari*). For more information, contact O'Reilly Media's institutional sales department: 800-998-9938 or *corporate@oreilly.com*.

Publisher: Roger Stewart
Production Editor: Nicholas Adams
Copyeditor: Octal Publishing, Inc.
Proofreader: Charles Roumeliotis
Indexer: Wendy Catalano
Interior Designer: David Futato
Cover Designer: Julie Cohen
Illustrator: Rebecca Demarest

August 2017:   First Edition

Revision History for the First Edition

2017-07-28:   First Release

See *http://oreilly.com/catalog/errata.csp?isbn=9781680450996* for release details.

Make:, Maker Shed, and Maker Faire are registered trademarks of Maker Media, Inc. The Maker Media logo is a trademark of Maker Media, Inc. *Getting Started with 3D Carving* and related trade dress are trademarks of Maker Media, Inc.

Many of the designations used by manufacturers and sellers to distinguish their products are claimed as trademarks. Where those designations appear in this book, and Maker Media, Inc. was aware of a trademark claim, the designations have been printed in caps or initial caps.

While every precaution has been taken in the preparation of this book, the publisher and authors assume no responsibility for errors or omissions, or for damages resulting from the use of the information contained herein.

978-1-680-45099-6

[TI]

Contents

Introduction.. vii
An Experiment.. xvii

1/Welcome to the Community.. 1
    The First Challenge... 1
    What Inspires You?.. 2

2/Getting Started.. 5
    Getting Started with Easel...................................... 5
    The Easel User Interface.. 6
    Engraving Your Tile.. 16

3/Inspiration Tile.. 19
    Meet the Maker.. 19
    Project 1: Engrave the Largest Digital Tile Wall in the World.. 20
    Skill Builder #1: Using Digital Calipers....................... 21
    Skill Builder #2: Picking Bit Diameters........................ 23
    Skill Builder #3: Selecting Carve Depth........................ 26
    Skill Builder #4: Clamping..................................... 28
    Your First Carve... 31

4/Glider.. 33
    Meet the Maker.. 33
    Project 2: Gliders! What Kind of Wings Fly the Farthest?...... 34
    Skill Builder #1: Tabs... 37
    Skill Builder #2: Double-Sided Tape............................ 38
    Carving and Flying.. 39

5/Inlay Cutting Boards.. 41
    Meet the Maker.. 41
    Project 3: Inlay Cutting Boards................................ 42
    Long Grain Style... 44
        Pick Your Shape.. 44

  Design the Inlay. 46
  Create the Carve Files. 47
 End Grain Style. 50
  Cutting the Wood. 50
  Bonding #1. 51
  Cutting Thickness Strips. 51
  Bonding #2. 52
  Sanding. 53
  Skill Builder #1: Designing for Your Bit Size. 53
  Skill Builder #2: Customizing Your Carving Calculations. 55
  Skill Builder #3: Clamping for Carving. 56
  Carve Away!. 57
  Cooking Up Some Epoxy. 58
  Sand, Sand, and Sand!. 58
  The Finish Line. 59
  It's Time to Eat!. 60
  Community Examples. 61

6/Fidget Spinner. 63
 Meet the Maker. 63
 Project 4: Fidget Spinner. 64
 Measure. 65
 Design the Spinner. 65
 Carve the Fidget Spinner. 69
 Sanding. 69
 Installing the Bearings. 70
  Speed!. 71
  Fidget Spinner App. 71
 Community Challenge. 72
  Community Examples. 72

7/Carving Out a Stamp. 77
 Meet the Maker. 77
 Project 5: Personalized DIY Greeting Cards. 78
 Background. 80
 Setup. 80
 The Artwork. 81
 Using the Stamp Maker App. 83
 Simulate Your Stamp. 84
 Creating Two Workpieces for One Project. 86
  Carving Sequence for Stamp Files. 87
 Finishing Techniques. 88

Using the Stamp... 89
Community Challenge... 89
    Community Examples.. 90

8/You've Started 3D Carving!.. 93
Where Did 3D Carving Come From?................................. 93
    Is CNC the Same as 3D Carving?.............................. 93
    Where Did the Term CNC Come From?........................... 95
    Before CNC It Was Called Card-a-matic....................... 96
    Say Hello to 3D Carving..................................... 97
Concepts and Lingo.. 98
    2D Versus 2.5D Versus 3D.................................... 99
    G-code.. 99
    Speeds and Feeds.. 101
    CAD, CAM, and Toolpaths..................................... 101
    Bit Descriptions.. 102
    Tool Material... 102
    Flutes.. 102
Early Access to 3D Profiling in Easel........................... 104
Where Can You Go from Here?..................................... 105

9/What's Next... 107
Jimmy DiResta... 107
    Box Joint Toolbox... 108
Warren Downes... 110
    Star Wars Chocolate Mold.................................... 111
Zach Darmanian-Harris... 112
    The Block Clock... 112
Bob Clagett... 113
    Four-in-a-Row Game.. 114
Casey Shea.. 114
    The Mighty Morphmobile...................................... 115
Steve Carmichael.. 116
    Making an Electric Guitar................................... 117
It's a Special Time in the World................................ 118

Index... 121

Introduction

Here at Inventables, we believe everyone is a maker. Every newborn is curious by nature, and toddlers explore the world by opening cabinets and dumping out pots and pans. Children let their imaginations run wild playing with things like LEGO and other construction toys. That's where making begins. These days, schools don't put an emphasis on exercising this part of your brain. Access to the internet and advancements in technology shift the focus away from learning information and towards gaining knowledge.

We also believe in the power of community. We believe a community that helps one another, especially when armed with easy-to-use tools, can learn together and change the world. This phenomenon has happened for centuries. Today, learning and change can happen faster than ever before in human history. We can watch videos, instant message, share photos, and answer questions to help one another. The goal of this book is to build a community of encouragement and make 3D carving accessible.

We are really excited about 3D carving. So far, Inventables has introduced free software (Easel), launched affordable X-Carve and Carvey machines, and donated 3D carving machines to schools, libraries, and makerspaces. However, we believe we need to do more. It's not enough to have access to equipment and software: people need to be inspired and connected to others who are on the same journey. Right now, millions of people don't have access to 3D carving. Our goal is to provide them access and bring them into our community. We hope this book inspires you, connects you with like-minded people, and helps you start your 3D carving journey.

Access means more than just carving. It means exposure to the design process; excercising your creativity and imagination; being inspired by projects other people have done; giving encouragement to others in the community; asking questions to the community; designing your own projects; going to a public makerspace to do a carve; and, in some cases, buying your own 3D carving machine.

This book will be successful if 100,000 people post a photo with the caption "#GSW3DCbook brought out the maker in me" on Instagram. Please post it only if you believe it.

What Will You Make with 3D Carving?

We believe that the best way to learn is to make something. This book shows you how to make five projects. In the process of making each project, you learn the techniques and can then apply them to your own projects. The photo below show the five projects that we'll make together.

You can see more community examples of these five projects at *https://www.inventables.com/challenges*.

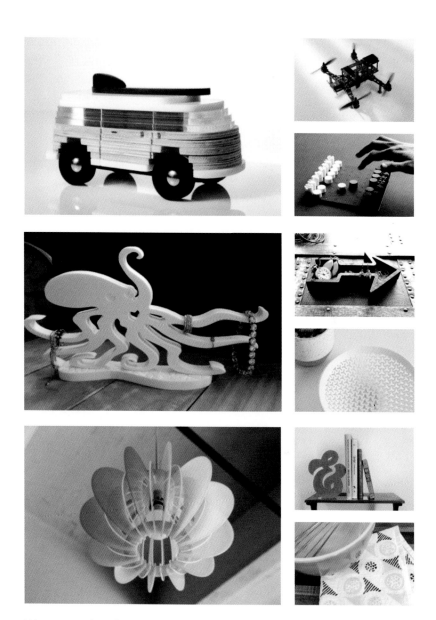

We want this book to inspire you. So, in addition to the five projects we'll work on together, we've shown you a few amazing creations from some of the greatest makers we know. On the cover of this book is an incredible 3D carved guitar by Steve Carmichael. In Chapter 9, you'll see projects from other great makers. Maybe you won't actually make your own guitar—that's a major undertaking—but seeing how it's done, reading the instructions, and watching the video on the Inventables website might inspire you to design your own *over-the-top* project. Greatness: that's what we want to inspire.

How You Can Get Started 3D Carving

For each project in the book, I'll outline everything you need. Of course, you'll need access to a 3D carving machine like X-Carve (Figure P-1) or Carvey (Figure P-2) in order to complete the project. If you don't have access to a machine, you can find them at many schools, libraries, and makerspaces near you. If you aren't able to locate one, go to the Inventables website forum. People there are quite friendly and they might carve it for you, or you might discover that there is someone who already has a machine that lives near you. Even if you don't have access to the machine, you can learn a lot from the design process. You can do everything except the carving until you eventually gain access to a machine. For safety reasons, children under the age of 18 should be supervised by an adult when using X-Carve or Carvey.

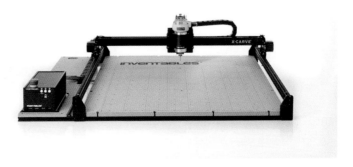

Figure P-1. *X-Carve by Inventables is a 3D carving machine designed to help hobbyists and small businesses get started*

Figure P-2. *Carvey by Inventables is a desktop 3D carving machine that contains sound and dust in an enclosed space*

We believe it is important for every school and every library in the world to have a 3D carving machine. Anyone who wants to learn about digital manufacturing—students, teachers, parents, and the general public—can exercise their creativity if they have access to machines. We are starting with a simple goal: by the end of 2020, we want a 3D carving machine in every school in the USA.

In 2015, we committed to President Obama and the White House that Inventables would donate a 3D carving machine to a school in every state. We delivered on that promise. This was our way of getting that process started in schools across the country. We still have a long way to go, and we need your help. If you are a teacher, parent, or student who wants a 3D carving machine but doesn't have the funding, please go to Donors Choose and start a campaign. Tweet the link @inventables and we'll help promote it as much as we can. If you get a 3D carving machine, let us know and we'll add you to the map at *http://www.inventables.com/50states*.

In addition to schools, we're working with libraries and camps to widen access in all communities.

Getting Started

Each project in this book assumes that you have access to an X-Carve or Carvey. It also assumes that you have completed the online step-by-step introductory guides for the machine(s) you are using. If you haven't, here's where you can find the guides:

- X-Carve (*http://easel.inventables.com/intro*)
- Carvey (*http://easel.inventables.com/carvey*)

If you don't have access to a machine, you can still enter one of our Maker Challenges digitally using our free design software, Easel. To submit your entry, go to *https://www.inventables.com/challenges*.

Getting Parts

There are two Make: kits that need to be purchased in order to complete the projects in this book: the Getting Started with 3D Carving Materials kit, and the Getting Started with 3D Carving Finishing and Bits kit. For teachers, each student will need a material kit, but one bit and finishing kit will work for every 10 to 15 students. If you don't have the kits, the full list of materials available for each project and can be purchased from Inventables (*https://www.inventables.com*) a la carte.

Conventions Used in This Book

The following typographical conventions are used in this book:

Italic
    Indicates new terms, URLs, email addresses, filenames, and file extensions.

`Constant width`
    Used for program listings, as well as within paragraphs to refer to program elements such as variable or function names, databases, data types, environment variables, statements, and keywords.

This element signifies a tip, suggestion, or general note.

 This element indicates a warning or caution.

Using Examples

This book is here to help you get your job done. In general, you may use the example projects in this book in your own exploration or curriculum. You do not need to contact us for permission unless you're reproducing a significant portion of the projects and reselling it. For example, writing a curriculum that uses several projects from this book does not require permission. Selling or distributing a book of examples from Make: books does require permission. Answering a question by citing this book and quoting example code does not require permission. However, incorporating a significant amount of example projects from this book into your product's documentation does require permission.

We appreciate, but do not require, attribution. An attribution usually includes the title, author, publisher, and ISBN. For example: "*Getting Started With 3D Carving* by Zach Kaplan (Maker Media). Copyright 2017, 978-1-680-45099-6."

If you feel your use of example projects falls outside fair use or the permission given here, feel free to contact us at *bookpermissions@makermedia.com*.

O'Reilly Safari

 Safari (formerly Safari Books Online) is a membership-based training and reference platform for enterprise, government, educators, and individuals.

Members have access to thousands of books, training videos, Learning Paths, interactive tutorials, and curated playlists from over 250 publishers, including O'Reilly Media, Harvard Business Review, Prentice Hall Professional, Addison-Wesley Professional, Microsoft Press, Sams, Que, Peachpit Press, Adobe, Focal Press, Cisco Press, John Wiley & Sons, Syngress, Morgan Kaufmann, IBM Redbooks, Packt, Adobe Press, FT Press, Apress, Manning, New Riders, McGraw-Hill, Jones & Bartlett, and Course Technology, among others.

For more information, please visit *http://oreilly.com/safari*.

How to Contact Us

Please address comments and questions concerning this book to the publisher:

> Maker Media, Inc.
> 1700 Montgomery Street, Suite 240
> San Francisco, CA 94111
> 877-306-6253 (in the United States or Canada)
> 707-639-1355 (international or local)

Make: unites, inspires, informs, and entertains a growing community of resourceful people who undertake amazing projects in their backyards, basements, and garages. Make: celebrates your right to tweak, hack, and bend any technology to your will. The Make: audience continues to be a growing culture and community that believes in bettering ourselves, our environment, our educational system—our entire world. This is much more than an audience, it's a worldwide movement that Make: is leading— we call it the Maker Movement.

To learn more about Make:, visit us at makezine.com. You can learn more about the company at the following websites:

> Maker Media: *makermedia.com*
> Maker Faire: *makerfaire.com*
> Maker Shed: *makershed.com*

If you have comments or questions about the book, please email us at *books@makermedia.com*.

Acknowledgments

I have a lot of people to acknowledge and thank. This book took a lifetime to write. I'd like to first thank my parents Larry and Carrie Kaplan for providing an environment that encouraged learning by discovery and asking why. This provided the foundation for everything I've achieved academically and professionally.

I'd like to thank my wife Allison, the most loving wife in the world, for giving me the space and time to run a company and somehow write this book at the same time. To my son Dylan, your curious nature and fearless enthusiasm is an inspiration.

There have been several people that played a significant influence on my life that specifically led to this book. My high school Sci-Tech teachers Jim Howie and Jeff Jordan who ignited the spark in digital manufacturing two decades before it was en vogue. Bart Dring, our Chief Engineer at Inventables, who designed MakerSlide and Carvey, and teaches our team with patience. Inventables software engineering team; Jeff Talbot, Paul Kaplan, Jim Rodovich, Ruwan Egoda Gamage, David Altenburg, Adam Hinz, and Eric Dobroveanu who build and improve Easel every day. Inventables' Marketing team: Alex Berger, Michael Fischer, Sam Alaimo, Mo Stych, Valerie Frank, and Nick Rissmeyer, who came up with the idea for Maker Challenges and put together the companion kits for this book. My assistants, Jules Woodard and Emily Michel, who somehow put up with me, my questions, and requests during the writing of this book. Finally, Edward Ford who reintroduced me to the field after being away for over a decade.

I've also gotten support in all sorts of ways from Phil Black from True Ventures who believed in me before it was crystal clear how Inventables would succeed. Eric Antonow, Richard Yoo, Morris Miller, Mike Hakimi, Bob Sabin, Georges Harik, Matt Mullenweg, Jameson Hsu, Joel Yarmon, Mark Hasbrock, Paul Bricault, Jon Callaghan, Ryan O'Nell, Sam Yagan, Steve Farsht, and Richard Lincoln for continued encouragement, ideas, support, and feedback as this journey unfolded.

You don't learn to walk by following rules. You learn by doing, and by falling over.
—Richard Branson

An Experiment

Traditional books are set up like traditional classrooms. The author is the teacher, and the readers are students. The author is the sage on the stage, and the reader is a consumer of knowledge. Readers are given information to digest and use it later when asked to prove what they learned.

That model works for many students, but it didn't work for me when I was in school.

With this book, I aim to be your guide on the side. You will participate directly in the process. You'll be the *maker of knowledge*. You'll read a little, do some experimentation, and get feedback from everyone else who is reading the book now or read it before you. Books could be interactive in the past. Today, we can use the internet to work together and learn from one another as a community.

I'm Trusting You

The premise of a traditional classroom is that students can't be trusted. After the teacher provides all the information, the student is required to do homework and study. Finally, a test is adminstered to measure how much knowledge the student retained.

There will be no test at the end of this book. The answers are not in the back of the book. I'm trusting that you will do your best work. In this book, you will be in control of your own learning.

Experiment #1: Sketch in This Book

The first project we're going to do together is draw a sketch using a pen. We're going to use a pen instead of a pencil because I want you to feel uncomfortable with the notion of drawing something permanent in this book. You can pick any color pen you want. In the space below, sketch something that you are interested in.

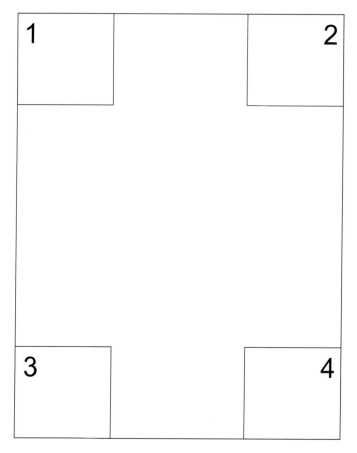

Now, you're going to create a secret code that only people who read this book will understand. Go back to your sketch and answer the following four questions in the four corners of the page.

1. Did you complete your sketch before turning the page? [Yes or No]
2. Use one word to explain why you did or didn't.
3. Did you first sketch on a piece of scratch paper before you committed to the sketch you put in this book? [Yes or No]

4. Use one word to illustrate the emotion you felt in the moments before you let the pen touch the page.

It's difficult to be honest about our emotions and put ourselves out there by sharing an idea. Putting the pen to a page in a textbook means you are taking a risk. There is a risk someone will judge your work. There is a risk your first sketch won't be as good as you imagined in your mind. You can't erase the sketch; it is now permanently on the page.

The experiment was tough for me, too. Here is my sketch and answers to the four questions:

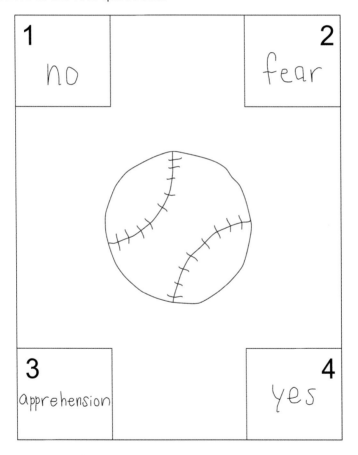

John Lasseter is the Chief Creative Officer of Pixar Animation Studios and Walt Disney Animation Studios. He's also the principal creative advisor for Walt Disney Imagineeering. In a 2006 interview with *USA Today*, Lasseter said, "Every single Pixar film, at one time or another, has been the worst movie ever put on film. But we know. We trust our process. We don't get scared and say, 'Oh, no, this film isn't working.'"

Here's the truth: I was guilty of letting fear hold me back while writing this book.

You probably expected this book to focus only on domain-specific technical lingo and techniques. In fact, when I started writing this book, it did focus on those things. After more than a year of work, I completed a full draft. Three editors from Maker Media reviewed it and shared feedback. Once I had a full draft finalized, I had the confidence to share it with a small group of employees at Inventables. I didn't feel comfortable sharing it with them until I had something substantial. I had a fear that I would be wasting their time. The day finally came for the book review meeting, four weeks before the deadline. I sat there anxiously awaiting their feedback. They were delicate with me, but in not so many words, they said it was all wrong. They said that it didn't inspire them. They said the projects weren't laid out well enough. They said if my goal was to get people started with 3D carving, the way I had written the book wouldn't do it.

I was happy that my team had the courage and insight to tell me this, but at the same time, I was sad that I had waited so long to ask for feedback.

For the last step in experiment #1, you're going to share your sketch and get some feedback. Grab your phone take a picture of the page in this book where you drew the sketch. If you are feeling brave, tweet your photo with #GSW3DCbook. By tagging it, you'll automatically group your sketch with everybody else's sketch.

Now, flip through the other tweets tagged with #GSW3DCbook and see how your sketch compares. Were other people scared to put pen to paper? How many people sketched in the book without a practice sketch? What emotions did other people feel? How does seeing everyone else's sketch make you feel now? Please don't publicly explain the secret code. We all took a risk and put ourselves out there. We are trusting you.

1/Welcome to the Community

We're all working on this 3D carving journey together. This book uses the mastery system model: you'll work on a skill until you master it. You have the opportunity to revise your work as frequently as you wish. Spacing out learning sessions makes you more likely to deeply process the information and remember it better and longer. You'll also have the opportunity to receive feedback from others in the community. The journey will be filled with success and failure, mistakes and learnings. We have young students and older students of life. I expect you to do your best work and be respectful of everyone else in the community.

For each project in this book, I worked with a maker from the community. They have created a project page on the Inventables website where you can find all the files, materials, and supplementary videos. You can also post your remake of the project. In addition, you can participate in our Maker Challenges so your work becomes a part of our global community.

We have three core values in the community: independence, collaboration, and kindness. In a traditional book, you'd read all about what those values mean. In this book, we'll do a project together as a community. My hope is you'll learn what these values mean through experiencing them. The project we are going to do together has never been done before. It's a project that, if we are successful, will set a world record.

The First Challenge

The Hanoi Ceramic Mosaic Mural (Hanoi Ceramic Road) is a ceramic mosaic mural on the wall of the dyke system of Hanoi, Vietnam. With a length of about 4 km and a total area of 6,950 m², it holds the record for the world's largest ceramic mosaic and was awarded a Guinness World Records certificate on October 5, 2010. The Hanoi Ceramic Road inspired us to think about tiles in a whole new way.

We need your help to create a digital tile wall that sets an all-new world record. To achieve the goal, our community needs to make 299,240 tiles. Each person will need to carve a tile featuring something that inspires them. Then, post a picture of the tile to our Mosaic Tile Maker Challenge to build the world's largest tile wall.

What Inspires You?

Draw a picture of something that inspires you in the space that follows. Keep the design out of the pink area. When we carve the tile, we'll need to put clamps in the pink area. (Later, we'll learn how to carve to the border of the material.)

Now, take a step back. Where does your inspiration come from? I asked people at Inventables where their inspiration comes from. Their answers fell into the following six categories:

- Problems
- Helping someone else or doing a project for someone else as a gift
- Seeing what other people are working on
- Seeing something out in the world
- An area or topic they are interested in
- Something that has never been done

Often, it's easier to come up with multiple designs for a project instead of always using your first design. By drawing a second tile, you'll be able to choose which design you'd like to carve in the next chapter.

Here's another blank tile in case you didn't like your other two tiles.

Now, using your phone, take a picture of each of the designs that you want to carve. Try to center it so that the picture is straight. In the next chapter, we'll load them into Easel.

2/Getting Started

The projects and skill builders in this book begin easy and increase in complexity. They're designed to build your confidence, experience, and success as you go. In addition, you'll learn that 3D carving is just one tool in the shop. For the more complex projects, you'll use other tools to finish the job. Along the way, I'll drop in small bits of knowledge, tips, and tricks. Since the projects were contributed by makers from the community, you'll get a chance to meet the maker at the beginning of each chapter.

3D carving is a journey. Think of it as if you were playing a sport or a musical instrument. With practice, you can build skills beyond what was possible in the beginning of your journey. Each time you practice, you get a little bit better. Focus more on the process than the outcome. With 3D carving, the journey is the destination.

Getting Started with Easel

In Chapter 1, you contributed to the largest tile wall ever created. In this chapter, you're going to load your mosaic tile design into Easel to prepare for your first carve. In Chapter 3, you're going to carve it out.

The first thing to do is get an Easel account, if you haven't done so already. Easel accounts are free. You can get one by going to *www.easel.com*. Then, click "Launch the app" to get to the login page.

If you have an Easel account, log in as usual. If not, select "I am a new customer." Type your email, full name, and a password of your choosing in the appropriate text boxes.

Easel is free web-based 3D carving software made by Inventables. Our software design and engineering team continually adds new features and functionality. Because it's delivered over the web, improvements are rolled out to all users immediately. This means that every time you log in, you are using the latest version. As a result, Easel's interface constantly changes. By the time you are reading this book, it is likely that Easel's interface has been improved. Use the diagram that follows as a reference, but don't be concerned if the version you see does not exactly match the version in this book.

The Easel User Interface

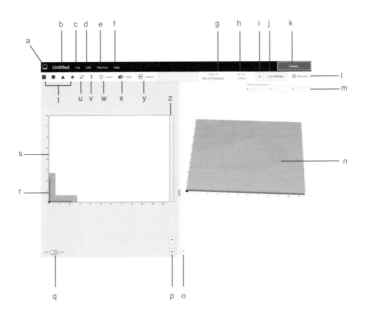

(a) **Easel button** This button opens and closes the project window where your files are stored.

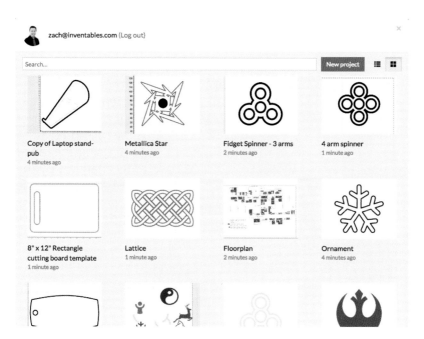

(b) **Project Name** This defaults to Untitled. Click it to rename your project.

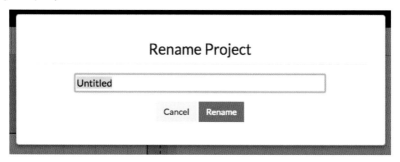

(c) **File menu** This menu contains commands relating to the handling of Easel files. You can create a new file, open an existing file, rename a file, or make a copy. You also can import files, change the public sharing permissions, and publish your project to Inventables' Project catalog.

(d) **Edit menu** Use this menu to edit information within the Easel file.

You can select all elements in the file, or select elements by cut type. The cut types available in Easel are outlines and fills. This feature is useful in a project like a puzzle: the artwork may be a fill, and the cuts to carve the individual pieces are outlines. You might want to set the depth shallow for all the fills, and all the way through the material for the outlines. The Edit menu lets you cut, copy, and paste whatever design elements are selected.

You also can use the Edit menu to manipulate the orientation of your project by using "Flip Horizontal" or "Flip Vertical." At the bottom of the menu, you can manipulate the layering of the elements (what's on top of what) by selecting "Send Backward," "Send to Back," "Bring Forward," and "Bring to Front." Changing the layering order of the elements changes what the carve will look like. Easel instructs the machine to carve from the front to the back (the same as carving from the top of your material to the bottom), so you can use these tools to set up the order of operations for your carve.

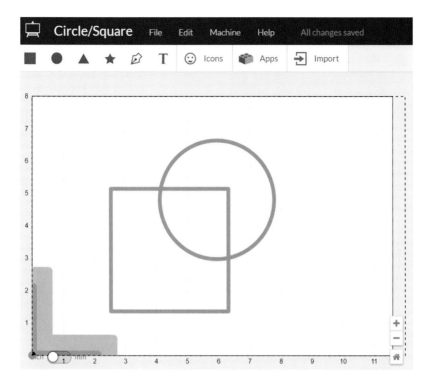

If two elements are selected, you can use the Combine feature. If the elements are overlapping, they will combine into one element (shown in the screenshot that follows). If they are not overlapping, they will become part of the same bounding box. However, the space between them will not be filled.

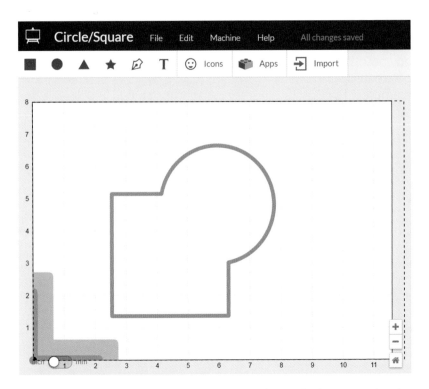

The last two features in the Edit menu do as their names imply: "Remove selected" removes the selection from the canvas, and "Center to material" takes the element and centers it on the material.

(e) **Machine menu** The Machine menu helps Easel communicate with your machine. The first drop-down menu prompts you to select the machine you're trying to connect. Select your 3D carving machine from the drop-down menu. If your machine connects successfully, the Carve button (k) will turn green. Carvey users can also "Home" or "Park" the machine from the Machine menu.

By clicking the Advanced menu, you can adjust the *safety height* and *stepover*. The safety height is the height your machine is above your material when it is not carving. You might want to adjust this to clear your clamps if they stick up more than 0.15 inches in the path of the rapid movements. Stepover is the amount that one toolpath is offset from another, represented as a percent of the toolpath's width. Think of it like mowing a lawn: you overlap a little bit of the last row when you turn around and go the other way. As a rule of thumb, the stepover should be between 10 percent and 40 percent of the carving bit's cutting diameter. For harder materials and 3D countours, lower the stepover to reduce the stress on the bit.

If for any reason you can't use the Easel driver to communicate with your machine, you can download the G-code for your file from the Advanced menu as well.

(f) **Help menu** This menu offers a number of ways to get help. The Community Forum is very active, so you can quickly get help from other Easel users. The Help Guides are written by the Inventables Customer Success team and provide more detailed information on each feature that is available in this book. You can report a problem directly to the Inventables Customer Success team, and they will respond on the same or next business day. From this same menu, you can download all versions—current and past—of the Easel driver.

(g) **Material menu** This menu lists materials for carving. When you know what material you are going to use, select it from this menu. When you make your selection, the material displayed in the 3D preview on the right side of the screen will change to the material you have selected. We are continually testing materials from the Inventables store and adding them to this menu.

(h) **Bit Size menu** The Bit Size menu lists all the bits that have been tested for the selected machine. Inventables uses a color coding system to make it easier to indentify and learn about the different sizes and styles of bits. If you are not using an Inventables bit, you can click "other" and add your own bit.

If you add your own bit, you will notice that the Cut Settings menu now displays a caution icon.

Inventables performs testing on all of the material and bit combinations presented in the menus to generate the recommended values for feed rate, plunge rate, and depth per pass. If you use your own bit, you need to calculate these values and enter them by clicking "Custom."

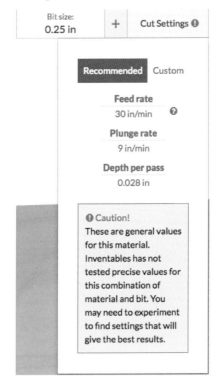

(i) **Additional Bit menu** As of this writing, this feature is a beta feature. It enables you to add a second bit for small, detailed areas that can't be carved with a larger bit. We'll go over this in Chapter 7.

Beta features are not in Easel by default. You can turn on beta features by going to Machine menu → Advanced and activating a feature you want to use. We are still testing these features and fixing issues before we are ready to release them to all users.

(j) **Cut Settings menu** This menu displays values for recommended feed rate, plunge rate, and depth per pass. Inventables did testing with each bit and each material to determine the recommended values for specific material and bit combinations. When you select a specific material and bit combination, the recommended values automatically update. If you want to type in your own values, click "Custom."

 The formula we used is Feed Rate = Spindle Speed × Number of Cutting Edges (flutes) × Chip Load (thickness of chips).

Feed rate is the velocity at which the bit advances along the workpiece. It is measured in distance per time. In Easel, we use inches per minute for imperial, and millimeters per minute for metric.

Plunge rate is the velocity at which the bit plunges or descends into the workpiece. It is also measured in distance per time. In Easel, we use inches per minute for imperial, and millimeters per minute for metric. For soft materials like foam, plunging can be done faster; for harder materials like metals, it is best to plunge more slowly.

Depth per pass is the depth the bit goes down each time it begins a new pass deeper into the material.

(k) **Carve button** This button turns green when the machine is connected. When you click the button, a walk-through is initiated that helps you confirm that everything is in order before you begin carving. (Think of it as a preflight checklist.) At the end of the walkthrough, click "Start Carving" to begin the carving process.

(l) **Simulate button** This button simulates the carve. It estimates how long the carve will take to complete and visualizes the path the tool will take during your carve.

(m) **Material dimensions** You can use these input fields to change the length, width, and thickness of your material.

(n) **3D preview** This displays a rendering of what your project will look like when it is completely carved.

(o) **Window adjustment** Drag this adjustment to the left or right to adjust the width of the window.

(p) **Canvas Zoom and Home buttons** The plus and minus buttons zoom the canvas in and out, respectively. The home button sets the size of the canvas to the full size available on the left side of the screen.

(q) **Units switch** This toggle switch changes the working units from inches to millimeters.

(r) **Smart Clamp™ keep out area** When Carvey is selected as the machine in the Machine menu, this red and gray area indicates the location of the Smart Clamp in Carvey's work area. The gray area is the physical clamp. The red area represents the area that must be clear in order to keep your clamp safe. Remember, the collet is wider than the bit: even if the bit avoids the clamp during a carve, your collet may not. This is why the red area must remain clear, too.

(s) **Canvas** This is the area in which you create, import, and edit your designs.

(t) **Shapes** The default shapes are square, circle, triangle, and star. You can add each shape to the canvas by clicking it. After a shape is on the canvas, you can drag it to move it around, resize it by dragging one of the gray anchor points, or edit the shape's geometry. To change a shape's geometry, double-click a gray anchor point. When the anchor point changes to white, you can drag it to create the geometry you want.

(u) **Pen tool** This tool makes it possible to draw a straight line and other shapes by using anchor points. The simplest geometry you can draw is a straight line between two anchor points. If you continue to click around the canvas, you will create a path made of straight lines. To stop adding path segments, double-click the most recent anchor point. By placing another anchor point on the first point, you can create a closed shape. After you are done creating your shape, you can convert straight-line segments to curves by double-clicking them while pressing the Shift key on your keyboard.

(v) **Text menu** Use this menu to add text or change the font to the text already in your design.

(w) **Icon menu** This menu provides access to hundreds of pieces of artwork in vector format. To add an icon to the canvas, simply click it. It can then carve as outlines or fills. Like shapes, you can double-click icons and edit them by moving or deleting the anchor points.

(x) **Apps button** This menu provides access to all the apps in Easel. At a high level, apps make it easier for users to make things. They do this by providing properties (e.g., sliders, input boxes, and drop-downs) that you can manipulate. Each time you change one of these properties, your app's executor function is invoked. Easel will provide the new values of all the properties, along with several other data elements representing the context of your project. Your app will use this data to create new design objects, or to modify existing ones. Example apps include the Inlay Generator, Box Maker, and Interlocker. The Easel API for creating apps is currently in early access. If you are a developer, we'd love your help to make it better. You can apply for early access at *http://developer.easel.com*.

(y) **Import menu** Use this to import SVG (scalable vector graphics) files or G-code. You can use "Image Trace" to generate a carvable vector file from a raster or image file.

(z) **Material boundary** This dotted line represents the perimeter of your material on the canvas. Any part of your design that you want to be carved must be within the boundary.

Engraving Your Tile

To start working on your title project, follow these steps:

1. When you first log in, you'll notice the project window is open. Start by clicking "Start a new project." You can also do this in the File menu (**c**) by clicking "New."
2. Name your project. Either click "Untitled" or click File → Rename and type the name of your tile project.
3. In the Material menu (**g**), select the type of material you are using. For this project, you'll be using two-color HDPE.
4. Next, click the Bit Size menu (**h**) and then select the blue 1/16″ 2FL Fish Tail.[1] Click anywhere in the lower area of the screen to close the Bit Size menu.
5. In the upper right of the window, in the Material Dimensions menu (**m**), type the X, Y, and Z dimensions. For this project, it will be X=6″, Y=6″, and Z=0.25″. We will be confirming the exact thickness in Chapter 3 with your digital calipers.
6. Click the Import menu (**y**).
7. Select "Image Trace."
8. Click "Upload file."
9. Click "Choose File."
10. Find where you stored the photo of your drawing (from Chapter 1), select it, and then click "Open."
11. Use the Threshold and Smoothing sliders to adjust your image until you are satisfied, and then click "Import."
12. Use the mouse to select your imported image.
13. Notice that the Cut/Shape panel appears when you select a shape. Click "Shape" to determine the exact size and position of the image. There are five radio buttons as well as X and Y coordinates. The radio buttons represent which part of the design the X and Y positions are being displayed for (upper-left corner, upper-right corner, center, lower-left corner, and lower-right corner).

---

[1] FL stands for flutes. Flutes are the cutting edges on the bit.

14. Select the radio button in the center and then enter 3 for the X position and 3 for the Y. This will put your design in the center of the 6" x 6" piece of material.
15. Go to the Size portion of the menu and click the lock icon. This will constrain the proportions of your design. Type the size that you want it to be. I suggest 3" or 4", whatever you think looks better. You want to leave room on the border for the clamps.

Now you're ready to move on to Chapter 3, where we will engrave your design into the tile.

3/Inspiration Tile

Meet the Maker

To finish up this first project, I teamed up with Jeff Solin (Figure 3-1), a computer science and making teacher, and FIRST Robotics coach at Lane Tech College Prep High School. He turned an old cafeteria into a full 4,000 square foot makerspace called the Innovation and Creation Lab. Jeff designed the curriculum for the Innovation and Creation Lab course and currently teaches more than 150 students. In his class, students learn art, engineering, math, physics, sciences, and programming. Students manipulate the physical and turn virtual ideas into reality. He uses Easel, X-Carve, and Carvey to show kids that there is a creative aspect to technology that they might not be used to seeing. You can see more about the space and student projects on Twitter at @LTMakers (*http://twitter.com/LTMakers*).

Figure 3-1. *Jeff Solin of Lane Tech High School in Chicago, Illinois*

Project 1: Engrave the Largest Digital Tile Wall in the World

The original inspiration for the tile project (Figure 3-2) came from Jeff Solin. He uses the project as a way to introduce his high school students to 3D carving. Each student carves their project in a 6″×6″×1/4″ piece of plywood or HDPE. These inexpensive materials are quick to carve, which allows 90 students to complete the project during class. To help deliver on the challenge from Chapter 2, you'll learn how to carve your drawing on a tile. Once it's carved, add it to our Maker Challenge and help us set a record. The Inventables team has made available a template project and companion video episode of Easel Live for this chapter at *https://www.inventables.com/projects/mosaic-tiles*.

Figure 3-2. *A table in Jeff's class with tiles made by his students in the Innovation and Creation Lab*

> # Shopping List
>
> *Tools*
> - X-Carve (*http://www.inventables.com/x-carve*) or Carvey (*http://www.inventables.com/carvey*)
> - A free account for Easel software (*http://www.easel.com*)
> - Digital calipers (*https://www.inventables.com/technologies/digital-calipers*)
> - A 1/16″ spiral upcut fish-tail end mill (*http://bit.ly/2tyLRxJ*)
>
> *Supplies*
> - One piece of 6″×6″×1/4″ two-color HDPE

In Chapter 2, we imported the drawing file and set the size and location. In this chapter, we will verify material thickness and bit selection, and set the carve depth. Next, we'll review proper clamping and carve. Finally, we'll upload your design to the challenge page.

Skill Builder #1: Using Digital Calipers

In Chapter 2, you entered in the nominal size of the material (.250″). The actual thickness is not always the same as the nominal thickness due to variation in the material, either from nature (for wood) or manufacturing processes (for plastics and metals). Grab your digital calipers (see Figure 3-3) to measure the exact thickness of the material. Close the jaws of the calipers and then turn them on. Then, press the zero button to ensure the starting position of the calipers is set to zero. Next, slowly open the calipers and place the jaws snugly on each side of the material. The display should provide a readout that is close to 0.25″, or twenty five hundredths of an inch. Recall that 0.001″ is one thousandth of an inch, and 0.010″ is ten thousandths of an inch. For reference, the thickness of a standard piece of printer paper is 0.003" (pronounced three-thousandths of an inch).

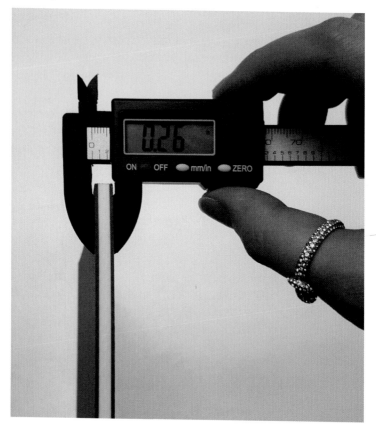

Figure 3-3. *Measuring the thickness of two-color HDPE with digital calipers*

 If your calipers use metric by default, press the mm/inch button to cycle through the options. Depending on the model, you might find a mode that displays inches in fractional units, as in 1/4".

You'll also need to know the overall material dimensions. Using a ruler, record the length and width of the material.

If you are planning to cut all the way through your material (which you won't be doing in this project), it is important to measure thickness precisely and set your Z-dimension appropriately. This is because the thickness reported by your supplier will not always be the exact thickness of the material. This is especially true with natural materials like wood.

If you fail to measure and set your Z-dimension incorrectly, you could end up not cutting deep enough or cutting too deep and into your waste board. You'll probably cut into the waste board a bit anyhow, but there's no reason to wreck it right away.

Skill Builder #2: Picking Bit Diameters

Notice on the design in Figure 3-4 that the word "Route" and the number "66" caused a warning to display on the right side of the 3D preview pane (Figure 3-5). The warning indicates the carving bit might not be right for the job (or vice versa).

Figure 3-4. *Problem areas*

Figure 3-5. *A warning that the bit is too large*

The red area on the 3D preview in Figure 3-4 shows the exact locations where the bit is too large in diameter to carve the design. Your design may or may not define an area that is smaller than your bit diameter.

To fix this situation, make sure that the bit diameter is less than the width of the narrowest elements in your design. As you make adjustments, watch as the error and the red area disappear (Figure 3-6) because now, the bit can completely carve out the design.

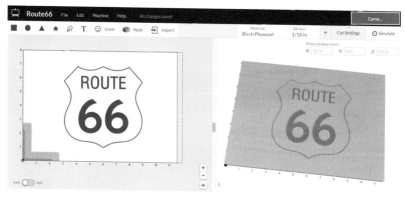

Figure 3-6. *Now the bit can carve without errors*

When selecting a bit, it is important to make sure that the diameter of the bit is smaller than the area it needs to carve. Figure 3-4 demonstrates how this problem often presents itself in sharp corners and with the serifs on letters. To achieve a high level of detail, you'll want to use an engraving bit that has a sharp point at the tip, such as the one shown in Figure 3-7.

Figure 3-7. *The bits with white and blue rings are engraving bits*

For most carvings, you want to select the largest bit diameter possible. It will take less time to carve with a 1/8" bit than with a 1/16" bit because you are able to remove more material with each pass. Remember the Additional Bit beta feature? This is where it comes in handy. You can use two different bits to complete one carve (see Figure 3-8 and Figure 3-9). Traditionally this is called a *tool change*. The idea here is that you can carve the majority of your design with a larger bit and then add a smaller bit to carve areas that are too narrow for the larger-diameter bit to access.

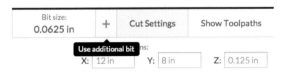

Figure 3-8. *The Additional Bit beta option*

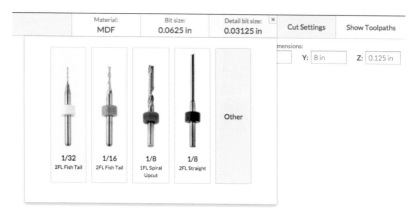

Figure 3-9. *Picking the additional bit*

Skill Builder #3: Selecting Carve Depth

There are two depth values to consider: *depth per pass* and *total depth of carve*. The depth per pass represents how deep the bit will plunge into your material for each pass it makes through the material. The total depth of carve represents how deep each design element will be carved into the material.

For example, if you need to carve a total depth of 0.120″ and your depth per pass is 0.030″, it will take four passes to completely carve the design. On prosumer-grade and consumer-grade machines like the X-Carve and Carvey, the machine can carve 0.120″ on a piece of foam in a single pass. On a harder material like wood or plastic, the machine is not powerful enough to carve that deep in one pass. Instead, it needs to make multiple passes. The Inventables team has tested each bit in the menu with each depth per pass and speed. To get the best results—which means nice clean cuts, minimal tearout, and minimizing broken bits—we suggest using the recomended values in the menu that we have calculated for you. If you are using your own bits or materials that are not listed, feel free to use your own values.

For the tile, we are going to engrave the design but not cut all the way through. Here's how to do this:

1. Select all the elements in your design.
2. In the Cut/Shape panel, click "Cut."
3. Change the depth to .060". This is the thickness of the top layer of color on the tile. We want to carve it off to reveal the second layer below.

4. Plug in the USB cord from your computer to Carvey or X-Carve and then click the Carve button.

 A broken bit is part of the 3D carving experience. It's one of the reasons that you—and anyone in the shop with you—must wear safety glasses when operating an open design 3D carving machine like the X-Carve.

Bits Will Break

You will invariably break bits, and you will break more bits when you are learning. If you forget to specify the correct material type, thickness, or type of bit in Easel, you run the risk of trying to cut too quickly (or too deeply), which will cause your bit to get very hot. If you have the incorrect settings with plastic materials, you might find that the plastic actually melts and fuses to the bit.

As you begin, it is important to develop good habits: always confirm your material type, material depth, and bit before carving. If you've set those correctly, you can leave it up to Easel to calculate the right speed and depth per pass.

Even still, you will eventually break bits. A 1/32" bit, although great for fine detail, will break faster than a 1/16" or 1/8" bit. To avoid wear and tear on your fine detail bits, be sure to use the additional bit option in Easel. It's more work to switch out the bit during the carve, but the carves will get done faster and your bits will last longer.

There is no "perfect" answer to setting the right depth. For any given set of variables, there is a limit to the performance you will experience from choosing depths at the ends of the range. In Easel, we have done testing to determine the recommended values. We did this testing to determine what produced good results on the Carvey and X-Carve. You can always click "Custom" and try your hand at modifying them. In Chapter 5, we experiment with this feature.

Skill Builder #4: Clamping

Clamping is often the culprit when something goes wrong. The reason clamping is so important is because the software and the machine assume that the workpiece is secure and not moving in any direction. If the workpiece begins moving during the cut, you'll begin to see some failure modes or unusual behavior. In this section, we're going to show you how to use the clamps that come with the X-Carve and Carvey machines.

The Carvey comes with the Smart Clamp hardware to secure materials to the waste board for cutting. The Smart Clamp is the black L-shaped clamp located at the lower-left corner of the carving table. You secure it using two hand-tightened screws. This clamp has a sensor with which your Carvey automatically zeroes your Z-axis to the top of whatever material you are using. Until you are an expert, it is vital that you use this Smart Clamp to secure your material on every carving job. If you do not use it, you risk damage to your machine.

Both Carvey and X-Carve come with clamps that use a triangle base and finger-clamp system, shown in Figure 3-10. These are traditional clamps that you might find on a Bridgeport mill. When using this style of clamp, you want to ensure that the teeth on the triangle base and the fingers are engaged with each other. The fingers should be at a slight downward angle, exerting force onto the material. If the teeth are not engaged or the clamp is parallel to the material, you risk the material slipping out from under the clamp during carving. You can never have too many clamps. On a Carvey, a good rule of thumb is to clamp with at least four total clamps (including the Smart Clamp). This means that you'll need five thumb screws because, as mentioned earlier, the Smart Clamp has two.

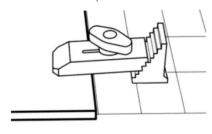

Figure 3-10. *The triangle finger clamp*

With materials thinner than 1/8", you might need more than four clamps. Sometimes, for very thin materials, you will also need some double-sided tape. As you carve away material, your workpiece begins looking like Swiss cheese. The missing material weakens the clamping force and your material tends to lift up. This can result in uneven carves or carves done at improper depths.

We also sell wood-bolt style clamps (Figure 3-11) on the Inventables website. Instead of the triangle base, the bolt serves as the adjustable portion of the clamp. Again, you want to have the clamp on a slight downward angle, so you are exerting pressure down onto your material.

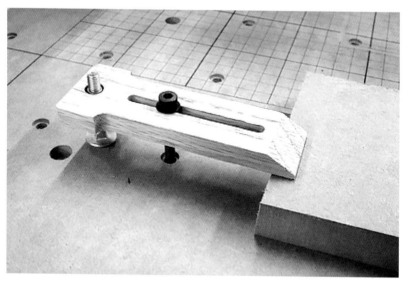

Figure 3-11. *Wood bolt clamps*

I'll add one more note about clamping: it matters a lot. Insufficient clamping is the silent killer on many carves. Eric Dobroveanu, one of the software engineers at Inventables, learned this through his experience machining aluminum. There's a significant difference in quality between good clamping and mounting the material to the waste board with screws. Instead of using the clamps we sell with the machine, he drills holes in the material and screws it down into the waste board. This makes it so the material doesn't move up, down, or in the X or Y directions.

You can use double-sided carpet tape to replace clamps when working with thin materials. It's awesome! It stays stuck when machining but peels up easily. This is especially useful as you begin to remove material because the remaining material will look like Swiss cheese and lose its rigidity as material is removed.

If you clamp poorly, the material will most certainly move. If you clamp well with the clamps provided, there will be more movement than if you drilled holes in the material and screwed them down, but potentially not enough movement to make a difference.

However, as you carve shapes out of your material, it loses rigidity and your clamps are less effective than when you started. You're no longer getting the benefit of the strength of the material because it's been carved away.

Your First Carve

Make sure that the USB cord is plugged into your computer. The Carve button in the Easel interface should be green. Click it and go through the wizard to confirm you have done each step correctly. When you get to the last wizard page, click the "Start Carving" button.

Congratulations! You started your first carve.

Please post your tile to Maker Challenge (*https://www.inventables.com/challenges*) and help our community beat the world record!

Don't forget to share a picture of your tile with the community on social media using the hashtag #inventables on Twitter, Facebook, and Instagram.

Welcome to the journey!

4/Glider

Meet the Maker

It doesn't matter what the material is, or what it's for: Bob Clagett loves making stuff (Figure 4-1). He enjoys showing other people how he works to inspire and empower them to make what they're passionate about. He went to art school and spent 16 years in the software industry before he jumped ship and began making stuff for a living. You can see more about his work at I Like to Make Stuff (*https://www.iliketomakestuff.com/*) and follow him on Twitter at @iliketomakestuf (*http://twitter.com/iliketomakestuf*).

Figure 4-1. *Bob Clagett of I Like to Make Stuff*

Project 2: Gliders! What Kind of Wings Fly the Farthest?

The original inspiration for the glider project came from Bob Clagett. He designed the project and did one in balsa wood (Figure 4-2) and one out of foam. In this chapter, we are going to focus on the balsa wood glider. Bob started his design in Adobe Illustrator. For this chapter, we're going to use his plane fuselage and design our own wings using the pen tool in Easel. Let's see who can get the most hang time!

Figure 4-2. *The balsa glider*

Bob's design ensures that the slits for the wings and tail are the right size. The height of these slits should match the thickness of the material used to make the wings and tail (see Figure 4-3). Bob saved the design as an *.svg* file, imported it into Easel, and then scaled them up to his final size. We're going to use part of his design in this chapter.

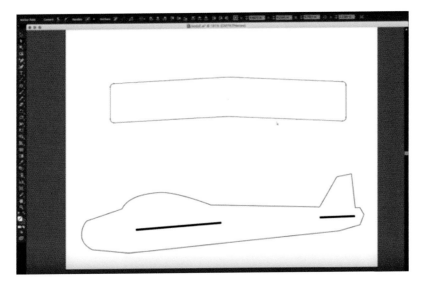

Figure 4-3. *Bob's original drawing in Adobe Illustrator*

Shopping List

Tools
- X-Carve (*http://www.inventables.com/x-carve*) or Carvey (*http://www.inventables.com/carvey*)
- A free account for Easel software (*http://www.easel.com*)
- Digital calipers (*https://www.inventables.com/technologies/digital-calipers*)
- A 1/16" spiral upcut fish tail end mill (*http://bit.ly/2tyLRxJ*)

Supplies
- One piece of 1/16"×24" balsa wood
- One piece of 3/32"×24" balsa wood
- A few paper clips
- Double-sided tape
- Markers or decorating method of choice

The first thing we are going to do is open Bob's file from the Inventables project catalog at *https://www.inventables.com/ projects/how-to-make-balsa-wood-gliders-using-easel*.

1. On the right side of the screen, click the "Open in Easel" button.

2. The Glider project opens as read-only, meaning you can't do anything with it. To be able to work on it, click the button in the upper-right corner, labeled: "Hey, you're playing in the Easel Sandbox. To save this project, make a copy."

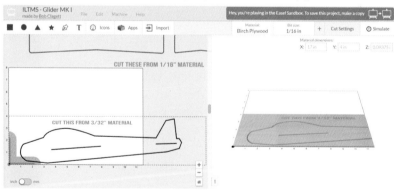

When you click this button, Easel makes a copy of Bob's file and places it into your Easel account. Rename the file by clicking the title and typing a name of your choosing.

3. In the Machine menu, select the type of machine that you'll be using. This will adjust your canvas to the right size for your machine. Bob was using an X-Carve. If you are using a Carvey, resize the fuselage—but not the slits—to fit on the canvas.

4. Click the slot in the fuselage of Bob's design and then go to the Shape tab on the Cut/Shape panel. Here, you can see the height and width of the bounding box. We need to determine the length of the line, so we'll use the Pythagorean theorem.

Length = $\sqrt{(4.250 \text{ in}^2 + 0.297 \text{ in}^2)}$

Length = 4.260 in

5. At the bottom-left of the screen, click on Workpieces. Workpieces allow you to keep all the pieces of a project together in the same file instead of using multiple files. Click the dropdown arrow in the first workpiece and select "Duplicate." This copies the current design into a new workpiece.

6. Name this workpiece accordingly for the wings. The wings use a thinner material than the fuselage. This helps the glider...well, glide. You can experiment with different thicknesses to see what works the best.

7. For practice, we are going to use the pen tool and trace Bob's wings. When you understand a bit more about the pen tool, delete Bob's wings and draw your own wings and tail. Get creative. Use trial and error to determine what types of wing shapes and angles fly the best. Keep in mind that they must fit in the slot, which measures 4.260″ wide.

Skill Builder #1: Tabs

When carving all the way through the material, it can be helpful to use tabs. Tabs are small bridges that hold the carved part to the stock material as the perimeter is carved out. They prevent the material from breaking loose and flying into the bit, damaging your new part. When you drag the cut depth to the full thickness of the material, the option for tabs appears on the Cut menu. Easel lets you set the number of tabs along with their height and width. I always try to use as few tabs as possible because it takes time to remove them.

Set the material dimensions for the length and width of the material you are using. Now, grab your digital calipers and measure the thickness. Enter all of these dimensions into Easel.

Next, select your wing. In the Cut/Shape pane, select the Cut tab and drag the cut slider down so you are carving through the full thickness of your material. At the bottom of the panel, make sure tabs are selected. For this design, two tabs is probably sufficent to hold the wing in place. You can move them by dragging them. I like to put them on the flat sides opposite each other.

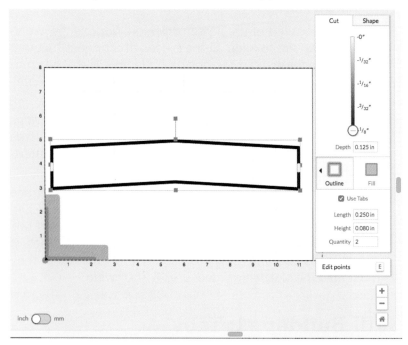

Skill Builder #2: Double-Sided Tape

Clamps are great for thicker materials. For thin materials, double-sided tape works well because it keeps the material flat and stable as sections are removed. If you use double-sided tape, you don't need to use tabs because the material will stick to the waste board when the perimeter is carved out. Keep in mind that balsa wood and expanded PVC are fairly fragile, so take extra caution when removing them from the tape.

If you are using a Carvey machine, you need to ensure that your material is underneath the Smart Clamp even if you are using double-sided tape. Recall that Carvey uses the lower-left corner as the zero point for the X and Y dimensions by default.

Carving and Flying

Make sure you have the 1/16" spiral upcut selected in the menu. Now that your material is secure, click "Carve" and carve your glider wing and tail from the 3/32" material using the Easel file. Next, cut the wing and tail pieces out of 1/16" balsa.

If you used tabs, use a utility knife to cut them loose.

Take some time to decorate your glider and make it your own. Sharpie markers or paint work great for this!

Now, slide the wing into the large slot in the body, and the tail into the smaller slot. You'll need to add a little weight to the nose of the glider. Two paper clips are enough to keep the nose balanced. Experiment with different weights, wing designs, materials, and material thicknesses to get the best flight.

Go outside and fly your glider!

When you get back to your computer, please post your glider to the Maker Challenge (*https://www.inventables.com/challenges*). Be sure to note how far it flew.

Don't forget to share a picture of your tile with the community on social media using the hashtag #inventables on Twitter, Facebook, and Instagram.

5/Inlay Cutting Boards

Meet the Maker

David Picciuto (Figure 5-1) is a full-time blogger and YouTuber who designs, creates, and teaches the art of woodworking and crafts. He has a deep passion for originality and design and truly believes that everyone can learn to be creative. Although he took some woodshop classes in high school, it wasn't until 2011 that he really began woodworking on his own. This new hobby became a deep passion for him and he made a career of teaching and sharing what he's learned. 3D carving allows him to incorporate his graphic design background into his projects. You can see more of his projects and videos at his website, *https://makesomething.tv*, where he shares his ideas and teachs others the beautiful craft of working with wood.

Figure 5-1. *David Picciuto of Make Something*

Project 3: Inlay Cutting Boards

There are two types of cutting boards: long grain and end grain. The terms "long" and "end" refer to the direction of the wood grain. long grain cutting boards (Figure 5-2) are much easier to make but harder on your knives. If you're getting started, we recommend making the long grain style. The end grain style (Figure 5-3) is easier on your knives and the colored inlay will add a bit of "spice" to your kitchen. However, they're more difficult to make unless you are an experienced woodworker and have access to a wood shop.

In this chapter, we show you how to make both types so you choose which is the right one for you. The long grain style was inspired by David's but made at Inventables. For the end grain style, David will walk you through each of the steps. To make the end grain style, you will need a bit of experience and knowledge of basic woodworking skills using the table saw and planer. We've posted the files and videos for each style online. The links are at the beginning of each of the following sections.

Figure 5-2. *A long grain cutting board made using the Inlay app in Easel*

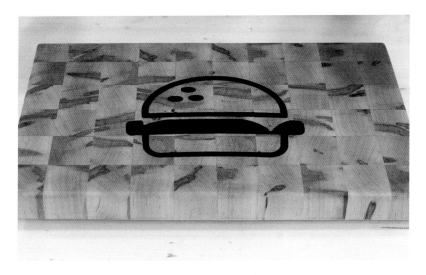

Figure 5-3. *The end grain cutting board*

In this project, you'll learn how to carve an inlay, install it, and add finish for a functional and beautiful cutting board for your kitchen.

> # Shopping List
>
> *Tools*
> - X-Carve (*https://www.inventables.com/technologies/x-carve*) or Carvey (*https://www.inventables.com/technologies/carvey*)
> - A free account for Easel software (*http://www.easel.com*)
> - 1/8" 2-flute straight-end mill
> - Clamps
> - Sandpaper or a random-orbit sander
>
> *Additional tools for the end grain style*
> - Table saw
> - Belt sander
> - Planer

Inlay Cutting Boards   43

Supplies for long grain style available at Inventables
- Maple hardwood 6"×12"×1/16"
- Walnut hardwood 6"×12"×1/2"
- Wood finishing kit (includes sanding block, 16-oz bottle of food-safe wood oil, and cotton application cloth)
- Wood glue

Supplies for end grain style
- 2" thick maple wood
- Epoxy
- Food-safe tint or dye
- Salad bowl finish
- Titebond II wood glue
- 6" wide Scor-Pal double-sided tape
- Rubber feet

Long Grain Style

The Inventables team has made available a template project and companion video episode of Easel Live for this chapter at *https://www.inventables.com/projects/6-x-12-cutting-board-template*.

Pick Your Shape

We've created three templates for the cutting board design. You can find the first one (Figure 5-4) by clicking the "Open in Easel" button on the aforementioned project page.

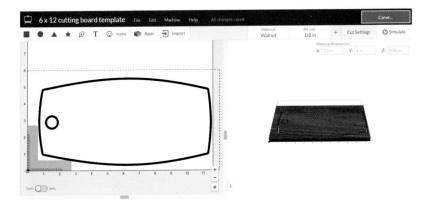

Figure 5-4. *Curved cutting board*

The other two templates (Figure 5-5 and Figure 5-6) are listed as links in step 2 on the web page.

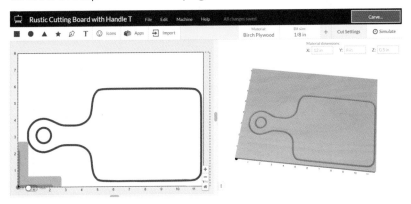

Figure 5-5. *Rustic cutting board*

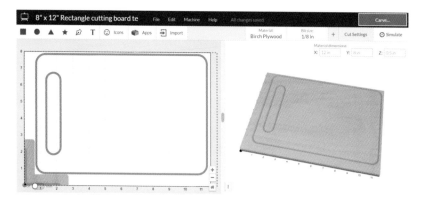

Figure 5-6. *Rectangle cutting board*

Design the Inlay

The next step is to draw or import the design for the inlay. You can choose a shape from the Icon menu, type some text, draw with the pen tool, or import an image from the internet, as illustrated in Figure 5-7.

Figure 5-7. *Design or import the inlay*

When you have the design you would like, select the design, click Apps, and then click the Inlay app (Figure 5-8).

The Inlay app will display both the pocket and insert part of the inlay. Choose the bit diameter you would like to use. The tolerance controls how tight the inlay will fit. If it is too tight, it will be very difficult to fit in. Making it too loose will show gaps between the insert and the pocket. We like to use a tolerance of .002″ for wood.

Click "Import" to add the design to the canvas.

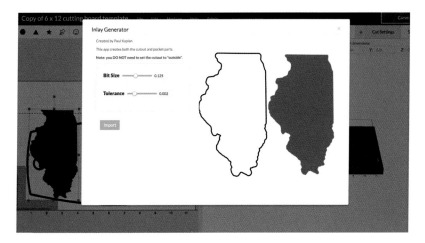

Figure 5-8. *The Inlay app*

Create the Carve Files

First, delete your original design so you only have the shapes imported from the Inlay app. Once imported, the pocket and the outline will be placed on the canvas near each other. The pocket is the solid-gray one. The material will be removed to a fixed depth. You'll want to ensure that the depth is the same as the thickness of the outline material: in this case, the material is 1/16" thick. The outline is the inlay insert of the design. You'll need to cut and paste the outline into a separate workpiece because it will be carved from the 1/16" maple hardwood.

Inlay Size

You must set the size of the inlay design before you use the Inlay app. If you resize the pocket and outline afterward, it will change the tolerance and the outline will not fit inside the pocket with the proper tolerance.

Make sure you have selected walnut as your material and you've set the dimensions to 6"×12"×1/2". For wood, you'll get good results with the 1/8" straight-cut mill. Straight-cut bits have vertical flutes that do not push material up or down. They are ideal for materials that can splinter easily, like woods.

Finally, position the remaining pocket design where you want it on your cutting board. If you want to place it in a precise part of the board, such as the center, use the Shape panel. First, select the perimeter of the board. Next, click the center radio button to find the position of the center of that element (Figure 5-9) and write it down. Click the pocket element. If you want the pocket in the center of the board, make sure its center is the same value as that of the perimeter.

Figure 5-9. *Positioning the pocket for the walnut board*

When you are happy with the position (see Figure 5-10), click the Carve button and carve out your walnut board.

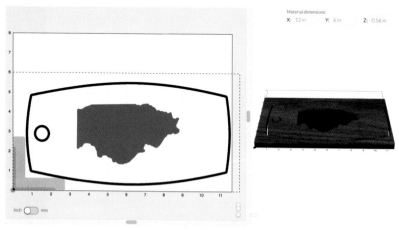

Figure 5-10. *Inlay pocket for the walnut board*

When the walnut board is finished, take it out of the machine. In Easel, load the second workpiece that has your outline (Figure 5-11). Make sure that maple is selected for the material and that the dimensions have been updated to 6"×12"×1/16".

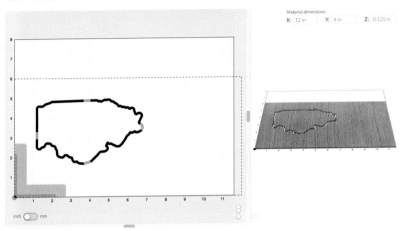

Figure 5-11. *The inlay insert*

For this carve, you'll want to use tabs or double-sided tape to hold down the material. The part will break free when the perimeter is completely carved out and it can fly into the bit, damaging it such that it won't fit into the pocket perfectly.

Because the 1/16" wood is so thin, you should be able to remove the tabs with a utility knife or scissors. Now, sand the insert and pocket lightly to remove any burrs on the edges. Test to ensure that the inlay will fit. If the fit is very tight, do not insert all the way before gluing or you won't be able to get it out.

Put some wood glue in the pocket and place the insert in it. To make sure the insert is as flush as possible with the surface of the pocket, you can either use another piece of wood on top of the insert and lightly tap it with a hammer or use the force of your hand to push it flush. When the insert is as flush as possible, wait for the glue to dry. Check the instructions on the bottle for the dry time.

After the glue is dry, sand the board to your liking. Use the mineral oil to give the board a shiny, finished look.

The mineral oil will keep the wood from cracking over time and is food-safe. You're now done with the long grain board and it should look like one of the examples in Figure 5-12.

Figure 5-12. *Long grain board examples*

End Grain Style

The end grain style cutting board is for the experienced woodworker. You should have access to a shop and experience using a table saw and planer, but you may new to 3D carving.

The files and companion video for the end grain style are located at *https://www.inventables.com/projects/cheeseburger-endgrain-cutting-board*.

The files were designed for the X-Carve. If you want to carve it on a Carvey machine, you'll need to resize it.

Cutting the Wood

You'll need some pretty thick wood for this project. I found some 2"×2" soft maple at my local hardwood dealer and planed it down to 1-3/4" thick. I then cut the wood into 1-3/4" thick squares at 20" long on my table saw, as shown in Figure 5-13. You'll need six pieces total for this cutting board. If you don't have the tools or the time to do this part of the process, some specialty hardwood or lumber stores sell pre-glued hardwoods.

Figure 5-13. *The prepared wood*

Bonding #1

Next up, bond the six pieces together in a random pattern using Titebond II wood glue and plenty of clamps. Slowly tighten your clamps, making sure your pieces stay aligned and don't slip out of place, as shown in Figure 5-14. Let this dry for six hours.

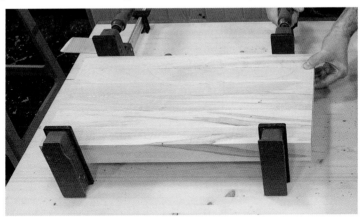

Figure 5-14. *Gluing the pieces*

Cutting Thickness Strips

Once the glue dries, take your work out of the clamps and run a couple of shallow passes through the planer to clean up the glue lines. You then crosscut 1-1/2" strips on the table saw (Figure 5-15). The width of the strips will determine the thickness of your cutting board.

Figure 5-15. *Cutting the strips*

Bonding #2

Flip all of your pieces so that the end grain is face-up, and bond everything a second time using Titebond II wood glue. Again, tighten the clamps slowly and make sure the pieces don't slip out of alignment (see Figure 5-16). Less slippage means less sanding and flattening in the next step.

Figure 5-16. *Bonding, round 2*

Sanding

Use a belt sander to flatten and clean up your cutting board (Figure 5-17). Use a very coarse grit paper and be prepared to sand for a while. Sanding end grain takes a long time. Patience is the key.

Figure 5-17. *Sand it flat*

Your material is now ready for carving. Before we can begin that part of the project, we need to work on the design. In this case, we are going to be making an inlay by embedding food-safe epoxy into a pocket in the shape of a hamburger carved into the wood. After it dries, we will sand it to be flush with its surface.

Skill Builder #1: Designing for Your Bit Size

I design most of my art in Adobe Illustrator. When drawing my artwork, I like to keep in mind what bit I'll be using in the 3D carving process. I know for this cutting board that I'll be using a 1/8" bit, so I want to confirm that all inside corners and radiuses are no larger than 1/16". That is the maximum size a 1/8" diameter bit can carve on an inside corner. A good example of an inside corner is the hamburger bun. On an outside corner, a round bit can carve a 90-degree angle. Notice in Figure 5-18 that there are no sharp edges on the inside because I'm using rounded corners.

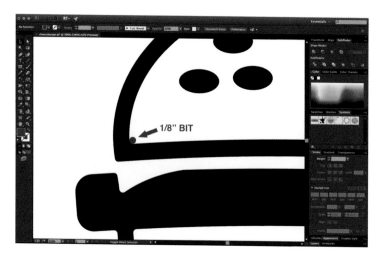

Figure 5-18. *Inside corner close-up*

After I'm happy with the design (Figure 5-19) I export the artwork in SVG format. For information on what settings you should use, read the SVG import guide online (*http://bit.ly/2t76qRC*).

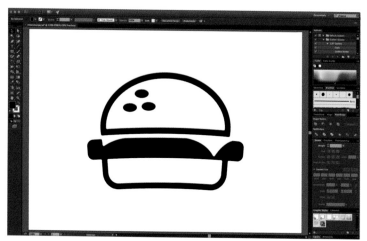

Figure 5-19. *The final hamburger artwork*

Skill Builder #2: Customizing Your Carving Calculations

Open Easel and log in with your Inventables credentials. Choose File → Import SVG and open your artwork (Figure 5-20). From here, you can set your material and dimensions. I'll visually align my artwork on the material. You can position the artwork precisely by using the Shape tab in the Cut/Shape panel. I'll be trimming the edges of my cutting board on the table saw in a later step, so exact placement isn't critical.

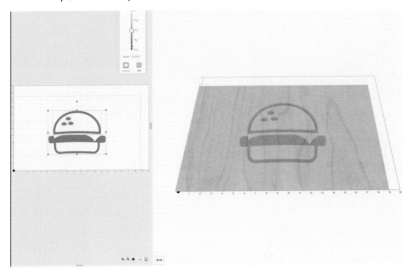

Figure 5-20. *The artwork in Easel*

When you select your material and bit, Easel automatically calculates the feed rate and depth per pass. You'll recall that the feed rate is the velocity at which the bit moves along the material. The depth per pass is how deep the bit goes into the material each time it carves your design.

When customizing this setting, I tend to gravitate toward a shallow depth per pass and a high feed rate, keeping in mind that different bits and materials will require different depths and rates. This tends to give a smoother finish. When customizing these settings, the formula to calculate your values is feed rate (IPM) = RPM x number of cutting edges x chip load. The *chip load* is the thickness of the chips. Chip load ranges are published by bit manufacturers for a given bit and type of material being carved.

The team at Inventables has tested the combinations of feed rate and depth per pass for each bit and material type in Easel. When you select the material and bit, Easel automatically selects the recommended feed rate. These are stored at the project-file level. If you are using a bit that is not in Easel, you can select "Other" and use the aforementioned formula to calculate your feed rate and depth per pass.

The big decision is determining how deep to carve the pocket that we will be filling in with epoxy. For an epoxy inlay, you don't need to cut any deeper than 1/4". In this case, we'll do about 1/8". To save a little time, you can set the total depth as a multiple of your depth per pass. For example if your depth per pass is 0.040", make your inlay .120" instead of .125" so that you can complete it in three passes instead of four passes.

Skill Builder #3: Clamping for Carving

There are a million ways to clamp your material down for 3D carving. My favorite way, and the easiest way in my opinion, is to use double-sided tape (Figure 5-21). I like to purchase the 6-inch-wide rolls. This is an excellent way to secure the material for projects that require carving all the way to the edge or on the sides of the material. I cut off a piece for each corner and a fifth piece in the middle to keep it secure. When you remove the tape at the end of the project, there might be some residue left over. You can sand it off later.

If you know you're going to be trimming off excess material from the edge of the project, you can also clamp or screw down your material right to the waste board. Both options are quick, easy, and don't get in the way of the milling bits.

Figure 5-21. *Use double-sided tape to help hold your material in place*

Carve Away!

After you have your artwork set up in Easel and your material secured in the machine, click "Carve" and watch the magic happen (Figure 5-22).

Figure 5-22. *Carving the design*

Cooking Up Some Epoxy

Now, it's time to fill the pocket we just carved with some food-safe epoxy. I've had luck getting some from Ebay. If you can't get your hands on food-safe epoxy, a two-part epoxy will work; however, you'll need to wait four to five weeks for it to fully cure before you can use your cutting board. There's a lot of conflicting articles on the internet regarding what is food-safe and what isn't. Typically, any finish is food-safe after it's fully cured and has a few weeks to off-gas. The polyurethane finish on your dining room table isn't food-safe when applied, but after it is fully cured, it's perfectly fine to eat from it. Do your research and choose what works best for you.

I like to tint my epoxy with dye or food coloring before filling the cavity. Also keep in mind that the epoxy will shrink and settle a bit, so you might want it to slightly overflow it so that it sits on top of the wood, as shown in Figure 5-23. This means that there will be some extra sanding in the next steps. Let this sit, dry, and settle for a couple of days.

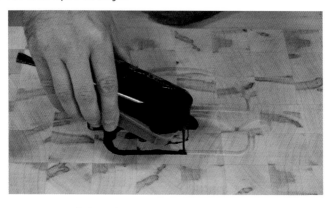

Figure 5-23. *Applying the epoxy*

Sand, Sand, and Sand!

After the epoxy is dry, it's time to trim off the edges of the cutting board using the table saw. Then, you'll need to sand and prep your cutting board for finishing. Begin with a very coarse grit of around 100 to level off the epoxy and smooth things out. Like I mentioned earlier, sanding end grain takes a long time. I also like to round over the edges with the random-orbit sander. Work your way up to 220 grit for a fine, smooth finish (Figure 5-24).

Figure 5-24. *Sand down the epoxy*

The Finish Line

For my cutting boards I like to use General Finishes Salad Bowl Finish. It's food-safe and leaves a protective film on your cutting board (see Figure 5-25). This type of finish needs less maintenance than a mineral oil/wax finish. Three coats over three days should do the trick.

Figure 5-25. *Finishing the board*

It's Time to Eat!

All that's left is to add some rubber feet to the bottom, demonstrated in Figure 5-26. Be sure to use stainless screws because they'll get wet when washing. Speaking of washing, you should only wash this cutting board by hand. Wood should *never* go in your automatic dishwasher.

Figure 5-26. *Mounting the rubber feet*

Figure 5-27. *Enjoy!*

> I hope you found inspiration in this project, and I encourage you to experiment with different woods and designs. Keep in mind that when choosing woods, it's best to stay with closed grain varieties because open grain could potentially trap bacteria. Be safe, stay passionate, and make something! —David

Please post your cutting board to the Cutting Board Maker Challenge at *https://www.inventables.com/challenges*.

As always, share a picture of your cutting board with the community on social media using the hashtag #inventables on Twitter, Facebook, and Instagram.

Community Examples

At Inventables, we're building an accessible community and tools for the maker journey. We hope that as you build your skills and confidence you are able to design your own remake of this project. Figures 5-28–5-30 show some similar projects other people in the community have created.

Figure 5-28. *Lousiana cutting board by Frank Graffagnino (https://www.inventables.com/projects/cutting-board-epoxy-inlay)*

Inlay Cutting Boards   61

Figure 5-29. *Cheese cutting board by John Walin (https://www.inventables.com/projects/remake-of-6-x-12-cutting-board-template)*

Figure 5-30. *Artists cutting board by Samantha Alaimo (https://www.inventables.com/projects/artist-palette-cutting-board)*

6/Fidget Spinner

Meet the Maker

Zach Kaplan (Figure 6-1) is the founder of Inventables and the primary author of this book. He holds a degree in mechanical engineering from the University of Illinois. Zach was first introduced to fidget spinners by 14-year-old Finn Callaghan. Finn, who earned seventh place at the Pacific Northwest Regional Yo-Yo Competition in 2016, designs and builds his own yo-yos and fidget spinners. In February 2017, Finn launched his own website called Alpine Spin Co. Finn first learned about spinners from YouTube, where he saw a video on how to make a spinner using popsicle sticks. This motivated him to get creative, and he started using his favorite material: duct tape. Now, he has a Carvey and is designing and selling his own fidget spinners. You can check out his YouTube channel (*https://www.youtube.com/spindle*).

Figure 6-1. *Zach Kaplan*

Project 4: Fidget Spinner

A fidget spinner (Figure 6-2) is a handheld toy meant for mindless fidgeting. You hold the toy between your finger and thumb and start spinning it. Almost everyone I show it to says, "I don't get it," until they begin playing with it. Then, they say "I get it now!" People of all ages have fun spinning these toys in their hands! But, in addition to being a fun toy, they also happen to be a good way to learn different carving techniques and design principles. The Inventables team has made available a template project and companion video episode of Easel Live for this chapter at *https://www.inventables.com/projects/fidget-spinners*.

Figure 6-2. *Fidget spinner*

In this project you'll learn how to make a press fit and install a bearing.

Shopping List

Tools
- X-Carve (*https://www.inventables.com/x-carve*) or Carvey (*https://www.inventables.com/carvey*)
- A free account for Easel software (*http://www.easel.com*)
- 1/8″ 1-spiral upcut end mill
- Clamps
- Digital calipers

Supplies available at Inventables
- Acrylic 8″ × 12″ × 1/4″
- Four 608-2RS bearings

Measure

You can find the files and companion video for the fidget spinner at *https://www.inventables.com/projects/fidget-spinners*.

The simplest fidget spinners use three bearings. As Figure 6-2 illustrates, the bearings are lined up in a row such that the two on either end spin around the bearing in the center. To achieve optimal spinning and a smooth feeling, the outer bearings need to be equidistant from the center point of the middle bearing.

To determine how far the outer bearings should be from center, you need to take a measurement of your hand. Touch your index finger to your thumb and then measure the distance between your hand and where your finger touches your thumb.

Design the Spinner

First things first: we need to design the spinner.

1. Change the units in Easel from inches to millimeters.
2. On the drawing tools menu bar, click the circle. This adds that shape to the canvas.

3. Select the circle and go to the Shape tab in the Cut/Shape menu. Make sure the lock icon is clicked so that Easel constrains the proportions of your circle as we resize it (so you don't end up making an oval).

4. The outer diameter of the bearings is 22 mm. This circle is going to be for the outer part of the spinner. Set the diameter of the circle to 40 mm to leave some material to hold the bearing (this leaves 9 mm of material on all sides of the bearing). If you are using a different bearing, use your digital calipers to measure the outer diameter.

5. With the circle still selected, click the center radio button in the Position options.

6. Set the position for the center of the circle to 40 mm in the X text box and 40 mm in the Y.

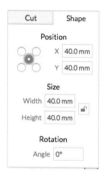

7. Adjust the settings for the material and bit. We are using acrylic material that is 8" x 12" x .250", and the 1/8" spiral upcut bit that has the gray indicator ring.

8. Copy and paste the circle twice so that you have three side by side (Figure 6-3).

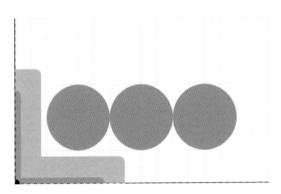

Figure 6-3. *Three circles side by side*

9. Subtract 5 mm from the measurement you did for your hand in the first section of this chapter, and note that measurement. This will provide a little clearance for the spinner to rotate freely; if you use the full length, the spinner would hit your hand.

10. Multiply this number by two in order to calculate the total length of the spinner from end to end. In my case, the measurement on my hand was 55 mm. So, when I do the calculation, I get: (55mm − 5 mm) x 2 = 100 mm. This means that the center of each circle should be offset by 30 mm. The distance between the centers of the outside circles is 60 mm, and the distance from the radius of each circle is 20 mm, for a total of 100 mm. If your hand is the same size as mine, set the center position of the second circle to X = 70 mm and Y = 40 mm. Set the center position of the third circle to X = 100 mm and Y = 40 mm.

11. Select all three circles at the same time. On the Edit menu, select "Combine" to make one shape (Figure 6-4).

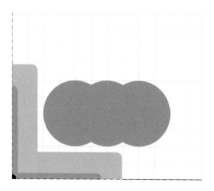

Figure 6-4. *Three circles combined*

12. Click the combined object and verify that the total width is 100 mm.
13. On the Edit menu, select Cut, click "Outline," and then select "Outside." This instructs the bit to cut outside the line.
14. Make the depth equal to the thickness of the material, which is 0.250″. Make sure tabs are unchecked.
15. Add another circle to the canvas. Select it, click the lock icon on the Shape tab, and then set the width to 22.3 mm.
16. On the Edit tab, select Cut, and then click "Outline" and "Inside." This will result in the inside of the circle being 22.3 mm. We're adding a tolerance of .3 mm because this is a *press fit* (also known as an *interference fit* or *friction fit*). This means that the bearing joins together with the acrylic without using glue, nails, screws, or other fasteners.
17. Copy and paste this circle two more times for a total of three, and set the center of each circle at the center of the other three circles. The final result will look like Figure 6-5.

Figure 6-5. *Our spinner ready for carving*

Carve the Fidget Spinner

For this carve, you'll want to make sure tabs are not checked and use clamps to hold down the material. If you are using a Carvey, it's possible for the part to break free when the perimeter is completely carved out. Be prepared to press the silver pause button to halt the machine if you see it break loose.

Clamp the material, load the bit, and then click "Carve." When the machine is done, go in with the vacuum and clean up the debris. Make sure the part is actually loose from the rest of your material before unclamping your stock material.

Sanding

If there are any imperfections, you can use a sanding sponge or piece of sand paper to clean up the edges of the spinner (Figure 6-6). Use a very coarse grit paper and sand slowly. Patience is the key. A little sanding goes a long way. If you go too fast, you might make it worse.

Figure 6-6. *Sanding the spinner*

Installing the Bearings

Place your newly carved part on a hard surface like a cement floor and see if you can push the bearings in with your hands. If you can't, that's OK. Grab a small piece of flat, soft wood and a hammer. With one hand, hold the wood on top of the bearings, applying even pressure. With the other hand, tap with the hammer a few times over each of the bearings. You should feel them lowering into the holes. Continue tapping until they are flush (Figure 6-7). A more enlightened way to press the bearings in is to use a watch press. You can order one from Amazon for about $15.

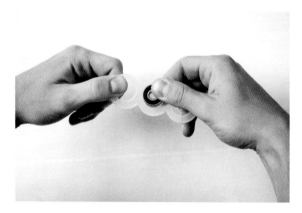

Figure 6-7. *Installing the bearings*

Speed!

The black plastic in the center of the bearing is called the *seal*. If you remove it you'll see the balls and some grease. To make the spinners go faster, you can remove the seals on the middle bearing by using a small screw driver or utility knife. This will reduce the friction. If you want to reduce the friction even further, you can spray a little bit of bike chain degreaser to remove all the grease. This will make it go faster: although the grease increases the lifespan of the bearings, it also adds friction. The rolling friction of metal on metal is lower than with the grease, thus allowing the fidget spinner to spin faster.

Fidget Spinner App

In the beginning of this chapter, we walked through how to design your own fidget spinner from scratch. The purpose of this was to help you understand a little bit about the design process. Fidget spinners have become quite popular at Inventables. One of our software engineers, Paul Kaplan, made an Easel app to make the process even easier. To use it, click into the App menu and select the Fidget Spinner app (Figure 6-8). With it, you can adjust the number of arms, their length, and the wall thickness. The app works with the same 22 mm bearings used in this chapter.

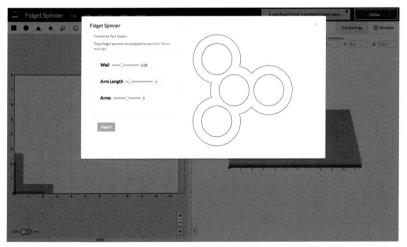

Figure 6-8. *The Fidget Spinner app*

I hope you found inspiration in this project, and I encourage you to experiment with different materials and designs. The sky is the limit with fidget spinner design and innovation. I've seen all kinds of geometries: four-bearing spinners, spinners with different sized bearings, and even some that have LEDs that light up when they spin!

Community Challenge

Contribute your fidget spinner design to the Maker Challenge (*https://www.inventables.com/challenges*), and together, we'll create a kind of toy store. Think of it as a virtual toy store. We want all the spinners to be unique and interesting, so try exercising your creative and technical muscles!

Please share a picture of your fidget spinner with the community on social media using the hashtag #inventables on Twitter, Facebook, and Instagram.

Community Examples

At Inventables, we're building an accessible community and tools for the maker journey. We hope that as you build your skills and confidence you are able to design your own remake of this project. Following are some projects from our community and the designers who made them (Figures 6-9 through 6-14).

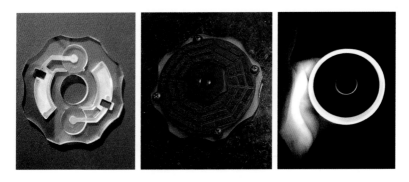

Figure 6-9. *Light-up spinner by Jason McDermott (http://bit.ly/2rBIZiC)*

Figure 6-10. *Jason McDermott*

Figure 6-11. *Black Widow by Finn Callaghan*

Figure 6-12. *Finn Callaghan*

Figure 6-13. *Wood Spinner (https://www.inventables.com/ projects/spinner-carvey-4) by Reece Hughes*

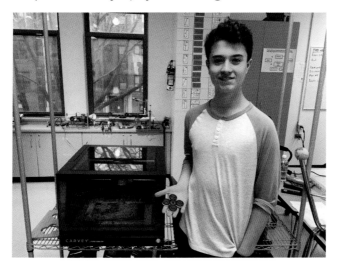

Figure 6-14. *Reece Hughes*

Fidget Spinner 75

7/Carving Out a Stamp

Meet the Maker

Travis Lucia (Figure 7-1) is a maker who focuses on 3D carving and woodworking projects. Travis was introduced to the maker movement and woodworking in 2011 when he purchased a set of basic power tools. He has since received formal training at the New England School of Architectural Woodworking. Quickly thereafter, he took over the family's two-car garage, filling it with tools and converting it into a fully functional woodshop. With his recent infatuation with woodworking and a long history of using computers and design, it was a natural evolution to marry these two passions with 3D carving. You can check out Travis's projects at *https://www.inventables.com/users/travis-lucia*.

Figure 7-1. *Travis Lucia*

Project 5: Personalized DIY Greeting Cards

We've all experienced it: you go to the store to buy a card for your friend or loved one and you leave thinking, "The cards at this store aren't very good." Cards at the store aren't very good because they aren't personal. The author of a card doesn't know *your* mom; he just knows moms. The cards at the store are intentionally designed to be generic enough to serve a broad market.

In this community project, we are going to create the most personal collection of DIY greeting cards in the world. You'll have the opportunity to create professional, quality stamps for use on a variety of materials, including paper, wood, and fabrics. I will then show you the process I use to print my stamped image on greeting cards and wood (Figure 7-2). 3D carved stamps are quick and fun, and the designs are limited only by your imagination. You can use the technique to make custom greeting cards, invitations, return address markers, art, and signs with all the artwork we create as a community.

In this chapter, we're going to carve out stamps from a piece of linoleum mounted on an Medium Density Fiberboard (MDF) block.

Figure 7-2. *Greeting cards*

Shopping List

Tools (Figure 7-3)
- A free account for Easel software (*http://www.easel.com*)
- Access to a 3D carving machine X-Carve (*https://www.inventables.com/x-carve*) or Carvey (*https://www.inventables.com/carvey*)
- Inventables 1/8″ single-flute upcut bit or a saw to cut out stamps (table saw, band saw, or circular saw)
- 1/32″ and/or 1/16″ milling bits (*https://www.inventables.com/technologies/upcut-fish-tail-spiral-bits*), depending on the level of detail in your design

Supplies
- Mounted linoleum stamping material 4″ x 6″ or larger (*https://www.inventables.com/technologies/linoleum*)
- A nail, awl, or toothbrush
- Ink stamp pad

Figure 7-3. *The tools you'll need*

Background

I had the opportunity to visit a local print shop that uses massive 100-year-old printing presses to create books, posters, and greeting cards. I was fascinated watching the process and appreciated the high-quality end product: raised ink and cardstock forming a card unlike any other. This inspired me to carve stamps and print my own greeting cards and invitations.

You can create stamps for so many different things, from adding a personal touch on a package to creating item tags for your crafts with your logo or name.

The Inventables team has made available a template project and companion video episode of Easel Live for this chapter at *https://www.inventables.com/projects/stamp-template*.

Setup

Open Easel and select linoleum as your material. Enter the dimensions of your linoleum block in the Material Dimensions fields. For this project, we will use an 8″ x 10″ piece, but you can modify it to use 4″ x 6″ or 5″ x 7″ in Carvey, and even 2″ x 3″ in X-Carve. The thickness of the linoleum is 0.100″, but we only want to engrave down 0.085″. The MDF is 0.75″ for a total thickness of 0.900″ in the Z-axis. The thickness of the linoleum and MDF can vary, so make sure to measure with your digital calipers.

 Using the 2″ x 3″ block in Carvey is not recommended as a beginner project because the Smart Clamp covers too much of the material. To do this successfully, you need to mount the linoleum without the Smart Clamp and set up a borderless carving operation. This is an advanced technique.

It helps to leave the stamping material oversized before carving your stamp. This approach leaves room for the clamps and for carving out the stamp.

The Artwork

Next, import your artwork into Easel or use the shapes, text, and icons provided by Easel to make a custom design. You can import an SVG file or create a vector of an image using the Image Trace feature.

A raster file is made up of lots of little pixels that combine to create an image. This is how your phone captures images when you take a picture. Unfortunately, a carving machine can't understand the information in a raster image. As such, we need to convert raster images into vector graphics. Vector graphics mathematically define the lines and objects that make up an image. Easel converts this information into instructions that direct the motion of the spindle on your machine. Figure 7-4 shows a sample design.

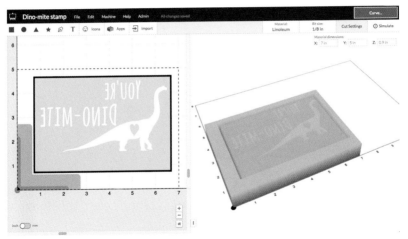

Figure 7-4. *The stamp design in Easel*

We used Easel's text feature and Image Trace to make the design in the example. Let's make your project even more personalized by drawing a design on white paper using a marker. Draw your own design and take a picture of it. It's important that the image has a very high contrast between what you want to show up and what you don't want to show up from your design. You should use only unlined paper so the lines don't show up in your design. Do your best to make sure the camera is parallel to the paper to avoid distorting the image. When you're satisfied with your photo, transfer it to your computer.

In Easel, click the Import button, and select "Image Trace."

In the dialog box that opens, click the "Upload file" button.

In the next dialog box, click "Choose file." Locate where you saved the picture on your computer, and then select the image.

Use the Threshold and Smoothing sliders to adjust the image the way you want it to look on the stamp. When you're ready, click the "Import" button." It's OK if there are some extra markings in the image—as long as they're not touching your design—because you can delete them after you import the photo.

 If your design contains very thin shapes or letters smaller than 1/4″, you might find that the linoleum tears away during the milling process, leaving you with an incomplete design.

Using the Stamp Maker App

In the previous projects, you've set the depth on a design element and then engraved or carved out the design. To create a stamp, it's just the opposite: you need to carve everything *but* the design. This technique is called creating a *relief*. The relief is raised above the carved background plane. Later, we will apply ink to the relief, which is the only part of the stamp that will make contact with the paper. This is how your design is transferred from the stamp to the paper.

To create the relief, you need to create a backer shape large enough to completely encompass the design you want to stamp, and a cut out shape for the border of your stamp. In the Easel App menu, there is an app called Stamp Maker that performs these steps for you. It reverses your artwork, raises it to the top, adds a backer shape, then adds a cutout border.

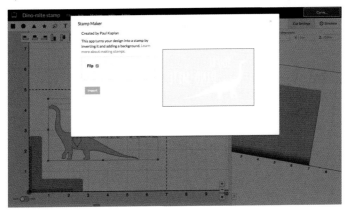

Carving Out a Stamp

Let's try the app on your project. Select your artwork, go to the app menu, and then click the Stamp Maker app. If you have text, you'll want to make sure the "Flip" checkbox is selected so your stamp imprints the mirror image of your design.

Flip ✓

The app sets the stamp design to a depth of zero and the back shape to a depth of 0.060". It will set the border (dark outline) to carve at a depth of 0.900", which is all the way through the linoleum block. You need to carve the stamp completely out or the raised areas will become part of the stamp design when you use it.

 You'll want to make sure that your design is away from the edges of the stamp so you have enough room to clamp it down on all sides. As a rule of thumb, I like to leave one inch around the border of my design for the clamps. On Carvey machines, you need to make sure your design is not under the Smart Clamp.

Simulate Your Stamp

In the upper-right corner of Easel, click the Simulate button. This shows the carving toolpaths (light blue lines) that your machine will make, as demonstrated in the following screenshot. Anything displaying as white is what will actually be the stamp. If you are satisfied with the level of detail, you're ready to move on. If you want more detail, you'll need to switch to a bit with a smaller diameter.

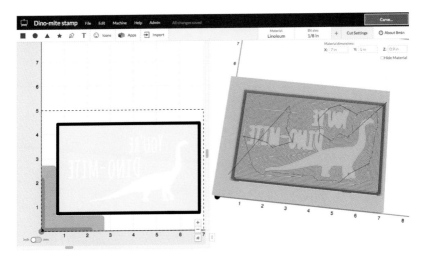

In the preceding screenshot, you can see that it's difficult to read the letters in the design. If we change the bit size to 1/16" or 1/32", there's a lot more detail on the stamp. In the following screenshot, I changed the bit to 1/16", which brought out enough detail.

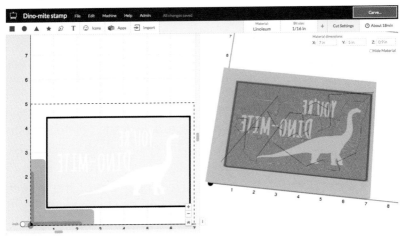

Notice that the file is set up to cut out the outline and the design. Unfortunately, we can't use the 1/16" bit to do both of these actions. The material is thicker than the length of the cutting edge of the 1/16" bit, as shown in Figure 7-5. This means that we'll need to use two bits to complete the project. The easiest way to do this is to create two different workpieces.

Figure 7-5. *The material is thicker than the bit*

Creating Two Workpieces for One Project

Make a copy of your first file by clicking into the Workpieces menu at the bottom-left corner of the Canvas. There is a small arrow in the upper-right corner of your current workpiece icon. Click the arrow and select "Duplicate." This will make an exact replica of your design; everything will be in the same spot when you carve it. In the original workpiece, select and delete the border. Now, for this workpiece, the machine will carve only your design (Figure 7-6).

In the second workpiece, delete everything *except* the border. Also, set your bit back to the 1/8" single-flute spiral upcut.

Figure 7-6. *Carving the stamp*

Carving Sequence for Stamp Files

Using the first workpiece (the one with the stamp), carve the relief with the smaller diameter bit. This will reveal your stamp design in the linoleum. After the file is done carving, use your vacuum to carefully remove the linoleum dust.

Next, change the bit to a 1/8″ single-flute spiral upcut. Close the first workpiece and open the second one (the one with the border). Select the border element. In the Cut tab on the Cut/Shape panel, adjust the depth to 0.9″, which will carve all the way through the material. Make sure you click the "Use Tabs" check box. This will leave small tabs in the material so your stamp doesn't come loose from the stock material. To detach the stamp from the stock material, remove the tabs using a knife or a chisel.

When changing the bit, you want to be as careful as possible to not move the workpiece or the spindle in the X or Y direction. It's important that everything stays where it is.

The new bit will be a different length. When using a Carvey machine, it will automatically calculate this change for you when it touches off on the button on the Smart Clamp. On an X-Carve, you will need to retouch off at your Z-axis zero position.

If the idea of using separate files and changing bits isn't something you want to do, you can use basic woodworking tools to complete the stamp. If you have access to a scroll saw, jig saw, or band saw, you can avoid the second file and bit change altogether. After the first file is carved, remove the material from the machine and cut out the profile of the stamp using the saw, as shown in Figure 7-7. If your stamp has straight edges, you could even use a table saw (which would be the fastest method).

Figure 7-7. *Cutting out the stamps*

Finishing Techniques

When your carve is finished, grab the shop vac and remove all the dust and chips. If you have lots of grooves in your design, small linoleum chips may be trapped in the nooks and crannies. Use an awl or a small nail to clean up any debris or dust that the vacuum can't remove. If necessary, you also can wipe the linoleum with a damp rag or paper towel to clean the surface.

Using the Stamp

Remove the lid from the ink pad. Lightly press your stamp into the ink pad a few times, making sure there is adequate ink coverage on the raised surfaces of your stamp. You may find it easier to press the ink pad onto the stamp, so you can see that the surface is covered. Avoid applying too much ink so it doesn't build up in the carved portions of your design. You can test your stamp on cardstock, cardboard, wood, and even fabric.

You will need to reapply ink to your stamp between each card to maintain the raised, individualized look (Figure 7-8).

Figure 7-8. *Applying ink to the stamp*

After using your stamps, be sure to clean the linoleum with soap and water to remove the ink and ensure that you have a clean surface every time.

Community Challenge

Contribute your stamp design to our Stamp Maker Challenge. Together, we'll create the most personalized DIY card collection in the world. We want a lot diverse artwork so that everyone is sure to find a unique and interesting design the next time they need a card. This will give you the opportunity to exercise your creative and technical muscles.

Add your design to the Maker Challenge at *https://www.inventables.com/challenges*.

Be sure to share a picture of your stamp with the community on social media using the hashtag #inventables on Twitter, Facebook, and Instagram.

Community Examples

At Inventables, we're building an accessible community and tools for the maker journey. We hope that as you build your skills and confidence, you are able to design your own remake of this project. Here are some similar projects that other people in the community have created (Figures 7-9 through 7-12).

Figure 7-9. *Trick-or-Treat bag by Steven Paxman (https://www.inventables.com/projects/rustic-3-color-block-printing)*

Figure 7-10. *Christmas stamps by Justin Robbins (https://www.inventables.com/projects/x-mas-gift-stamps)*

Figure 7-11. *Valentine's owl by Valerie Frank (https://www.inventables.com/projects/valentine-owl-stamp)*

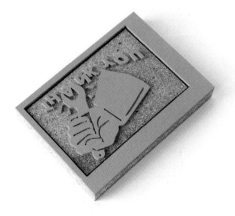

Figure 7-12. *Thank you stamp by Nick Rissmeyer*

8/You've Started 3D Carving!

Congratulations, you've ventured into something completely new! The first section of this book was designed to get you started as quickly as possible. We took you on a journey from sketching to engraving to carving.

After I get my hands dirty and make progress on learning something new, I'm interested in learning about the history of the field, the people who are involved, and what other resources are available. I can appreciate how talented or experienced others are, and this motivates me to continue learning. This chapter will outline a little bit more about the field to help you understand where all of this came from, introduce you to some of the language used in 3D carving, and offer some ideas on where you can go from here.

Where Did 3D Carving Come From?

Thanks in part to all the media attention lavished on it, most people have heard of 3D printing. And thanks in part to its proliferation in libraries and schools, quite a few people have actually seen 3D printing in action. If you're familiar with 3D printing, the term 3D carving might be puzzling. What can you do by carving away at material that you *can't* do by adding it one layer at a time? Turns out, quite a lot.

Is CNC the Same as 3D Carving?

Although many people have heard of or seen 3D printers, a much smaller number of them are familiar with the term *CNC machine*. (An even smaller number of people have heard of *NC* machines.) As it turns out, 3D printers and 3D carving machines are very specific kinds of CNC machines. But more often than not, when someone refers to a CNC machine, they are talking about a machine that cuts away at material using a rotating router or milling bit.

CNC stands for *Computer Numerical Control* and NC stands for *Numerical Control*. The concept of controlling a machine numerically is generally credited to a gentleman from Michigan named John T. Parsons (*http://bit.ly/2rDyF9x*) and his father's company, Parsons Corporation. If you dig into the history, you might hear that the first NC machine tool was demonstrated at the Massachusetts Institute of Technology in 1952. In actuality, both are true, but John never attended MIT—or any college.

John Parsons was a maker. He began working at his father's company as an apprentice in the tool room when he was 14 years old. By the time he was 20, he'd risen through the ranks and became general manager of the automotive division. As the United States became increasingly involved in the war effort during World War II, Parsons was of the age to enlist, but he was unable to be a soldier due to a physical condition. Nonetheless, he wanted to do his part. He responded to an advertisement to make parts for land mines and he won the contract because he was the lowest bidder. Instantly, he thought he screwed up. It turns out he bid the project correctly, he was just approaching production more efficiently than his competitors.

Two years later, in 1942, Parsons met a man named Bill Stout, the chief engineer for the Ford Tri-Motor airplane. In one of their conversations, Stout told Parsons that it was not possible to have an airplane take off or land at speeds slower than 50 mph. Shortly afterward, an article in the newspaper described a new kind of aircraft that did just that: a helicopter.

Parsons called Stout to challenge his earlier claim. Stout explained that a helicopter was an entirely different type of craft and urged Parsons to check it out.

Parsons contacted Sikorsky, the company that made the helicopter, and flew up to meet with representatives. At this time, Parsons had no aerospace experience aside from the time his father had made a wooden propeller for a stunt pilot. He explained that to the purchasing agent, but said he was interested in making parts for the company because his military contract had just been cancelled.

Two weeks went by, and Parsons had not heard anything, so he called to follow up. The purchasing agent said that due to his aerospace experience, the company decided to award him the contract to make propeller blades! Keep in mind that, for the previous few years, Parsons's company had been making land mines and bomb casings. His aerospace experience experience didn't add up to much.

> ## Bias Toward Action
>
> This was a case of having a *bias toward action* and having the confidence to figure things out as you go. Remember, Parsons never went to college. The approach his company took was similar to the way two of my mentors, Jim Howie and Jeff Jordan, taught *my* Sci-Tech class at Glenbrook North High School. They sought out complex problems for which a solution was not straightforward. With a problem in hand and knowledge of the tools available, they would begin figuring out how to solve the problem and create whatever it was they were trying to make. With a little bit of design, a little bit of engineering, and a lot of trial and error, they figured it out.
>
> Often, the greatest source of motivation and learning is getting in there with a real deadline and figuring out how to do it. That's really what this book is all about: a transition from the theoretical to the practical. A move from inaccessible to accessible. A journey from a world of industry experts that know it all to a world with a beginner's mindset. A move from CNC to 3D carving.

Where Did the Term CNC Come From?

In 1944, Parsons won the contract to make the rotors. At the time, machining parts was handled manually by a professional machinist in the lab. Parsons and his team decided they would make fixtures to streamline the process and make it more accurate. The rotor blades were built with a step-tapered tubular steel spar with plywood ribs and fabric skin. The metal was spot-welded to the aircraft structure. The technique worked out great, but one of the first 18 blades failed and the pilot was killed.

Parsons had to come up with a new solution—and fast. He convinced a few of his employees to come in on the weekend and work on a solution, which he presented the next day to Sikorsky. The presentation worked, and Sikorsky was back on board. Impressed by his bias towards action, now they were interested in hearing his ideas about metal blades. He had a background in stampings, which gave him an idea on how to approach the problem from a different perspective. Instead of making the blades with a bunch of pieces laminated together with the fabric skin, he proposed stamping out a single piece. The problem was that making a die in the exact shape of the airfoil would be near impossible because of its complex curve. Parsons asked one of the engineers to apply his college education and identify 200 points along the radius of each surface. This generated a chart describing the X and Y axis coordinates for a milling machine. Think of it as the earliest form of G-code.

To get the machine to precisely carve the die, they needed servo motors that could hold accurate positions. Modern stepper motors such as those on the Carvey and X-Carve machines weren't first described until 1957 and didn't go into production until the 1960s. So, Parsons went to the MIT Servomechanisms Laboratory. The Servomechanisms Laboratory was established at MIT in 1940 under the direction of Gordon S. Brown, then assistant professor of electrical engineering. During World War II, the laboratory's teams of research scientists and graduate students did research and development of servo motors and feedback control systems for the US Navy's gun-positioning instruments. Brown's team was at the forefront of the technology.

Before CNC It Was Called Card-a-matic

At this point, Parsons had the designs for what he called a Card-a-matic Milling Machine. How did it get this odd name? The machine was designed to use an IBM punch card reader that fed cards to an IBM calculating machine like the IBM 024 or the IBM 026. Thus, these machines were first called Card-a-matics!

Parsons's company hosted a contest at MIT to pick a name for the machine. (I guess Card-a-matic didn't sit well with everyone.) In the end, "Numerical Control" was the name that was chosen, and the winner received a $50 prize. With that little naming contest at MIT, for the next 65 years, makers referred to these machines as NC machines. Later, as punch cards were replaced by the computers we know today, the "C" was added; now, we have the abbreviation CNC. In Parsons's days, the primary machines they controlled were milling machines. As a result, the acronym stuck. When someone says CNC, even today, it generally means a milling machine controlled by a computer.

It's actually somewhat archaic to apply the term CNC only to milling machines. Technology has come a long way, and we are fortunate enough to have lots of CNC machines. Technically, a 3D printer, laser cutter, cutting machine, and vinyl cutter are all "CNC" machines. In the same way that NC was prefixed with "C" when the computer was added, it's time to dispense with the subdivision of the entire CNC category.

Say Hello to 3D Carving

3D carving is a more descriptive name for what the winner of the MIT contest called Numerical Control and what Parsons called the Card-a-matic Milling Machine. We say 3D because it can be used to create three-dimensional surfaces. We say carving because the machine operates by carving away layer by layer until the desired geometry is achieved. Figure 8-1 shows the process in action.

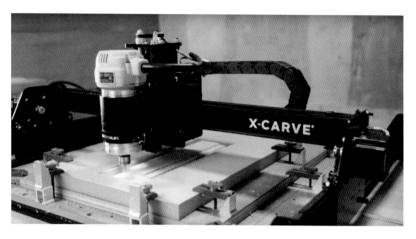

Figure 8-1. *Carving a 3D part*

3D carving machines are exciting because they allow you to make quality objects out of a variety of materials, including wood, metal, and plastic. Although these machines were first developed in response to military contracts, private companies and research universities each had their hand in advancing the technology. The machines started out quite expensive, such that only the military had access to their output. Over the past 65 years, they have progressively and dramatically fallen in price.

As we move into an era when these machines are going to be used by a larger group of people, we need a name that is easier to understand and more descriptive of what they do. From this point forward, I'll use the term "3D carving machine" instead of "CNC machine." I hope John Parsons would be pleased.

Concepts and Lingo

As you get deeper into any subject or field, you begin coming across words that seem strange or foreign. People who are familiar with these words say them casually and expect you to know what they are talking about. When those words are new to you, it feels like they are speaking in some secret language. In this section, we are going to explain some of these words and concepts.

2D Versus 2.5D Versus 3D

Sometimes, you might hear someone say they are making a 2D carve or 2.5D carve. You might be wondering what they are talking about. When you are working in two dimensions (2D) it means that the features on your part are all at the same depth. Two and a half dimensions (2.5D) means that you are cutting a part that has multiple flat features at varying depths. Three dimensions (3D) means you have a curve in your design. Figure 8-2 shows 2D, 2.5D, and 3D geometries from left to right.

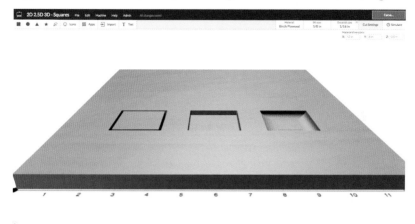

Figure 8-2. *2D, 2.5D, and 3D parts*

G-code

G-code is the code and programming language used to send instructions to the machine. With a program like Easel, you don't write the G-code by hand; Easel generates it automatically for you. In high school, my first project was to write my own G-code. During this process, I learned the meaning of the different codes. For example, G21 informs the machine that the values are in millimeters (mm), and G1 sets a position to which the bit is instructed to go. Writing G-code by hand is an educational exercise to understand how it works. However, complex parts that have curves could have millions of lines of code, so it's not practical to write your own code. Following is an exerpt from an Easel file:

```
G90
G21
G28
G21 G38.2 Z-71 F80
G10 L20 P1 X17 Y-12.25 Z12.7
G54
G0 Z20
G0 X20 Y20
M3 S12000
G0 X153.850 Y86.114
G1 Z-1.270 F228.6
G1 X155.439 Y86.351 F609.6
G1 X156.964 Y86.732 F609.6
G1 X157.491 Y86.906 F609.6
G1 X147.340 Y86.906 F609.6
G1 X148.666 Y86.502 F609.6
G1 X149.306 Y86.358 F609.6
G1 X150.950 Y86.114 F609.6
G1 X151.586 Y86.068 F609.6
```

If you want to export your G-code from Easel to see what it looks like, go to the Machine menu, and then click Advanced. In the Advanced Settings dialog box that opens, click the "Generate G-code" button. Once the G-code is generated, you can export the file to your computer. Open the G-code file using a text editing program on your computer to change the raw G-code. Just make sure you know what you are changing!

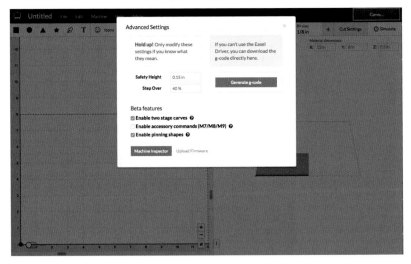

Speeds and Feeds

When people talk about speeds and feeds casually, they generally mean all the settings in the software instructing the machine how fast to carve your part. Technically, "speeds" refers to the cutting speed or rotational velocity of your spindle and, thus, the cutting bit. This is usually measured in revolutions per minute (RPM). "Feeds" refers to the feed rate or linear velocity at which the bit is moving as it carves. This is typically measured in inches per minute.

In addition to RPM and feed rate, people refer to additional settings like depth per pass, plunge rate, safety height, stepover, and profile stepover. Let's look at each one:

Depth per pass
> This is the distance in the vertical (Z) direction that the bit plunges downward into your material each time it begins a new layer of carving.

Plunge rate
> This is the velocity at which the bit travels in the vertical (Z) direction when it plunges downard into your material.

Safety height
> This is the height above your material that your bit moves to when it is not carving.

Stepover
> This is the offset distance in the X or Y direction that the bit moves over as it is carving your design. It's similar to the overlap that occurs when you're mowing your lawn: there needs to be a little bit of overlap between the last row you moved and the next row or you will have a little line of grass sticking up. The same principle is true with carving. In our case, the stepover is a percentage of the diameter of the bit.

Profile stepover
> This is the offset distance in the Z direction.

CAD, CAM, and Toolpaths

Before Easel existed, 3D carving files had to be created using computer-aided design (CAD) software. You would import the

CAD design into computer-aided manufacturing (CAM) software. The CAM software takes your design and creates the path that the tool bit will travel to carve your object. The *toolpath* is the path on which your bit travels to carve out your design.

Bit Descriptions

Your thinking around bits can start simple and expand over time. The basic parameters of a bit are the material it is made from, its shape, and its diameter. With each parameter, there is a lot of science and engineering behind how to select the right bit. In Easel, we have simplified the process by testing lots of bits and providing a simplified selection along with all the settings required for certain materials. When you are ready to learn more about bits, you'll need additional resources beyond this book. At Inventables, we sell Onsrud bits. The company has a very comprehensive production cutting tool catalog (*http://bit.ly/2t7C5Cr*) that is a great reference.

The next section presents a quick summary from their catalog.

Tool Material

Solid carbide
    Most bits are made from this material. The material provides the best rigidity and a long tool life.

Carbide-tipped
    This is a hybrid bit. The body is high speed steel but the tip is made of carbide. This gives the bit the wear resistance of carbide and the toughness of a high-speed steel (HSS) body.

High-speed steel (HSS)
    This type of bit is primarily used in hand routing. The material provides a tough body and sharper cutting edge, so it is a good bit for 3D carving.

Flutes

The flutes are the cutting edges on the bit. Flutes come in different shapes, and bits can have different numbers of flutes.

Flute shapes

Straight flute
　These have vertical flutes that do not pull material up or down. They are ideal for materials that split easily, like plywood and bamboo.

Upcut flute
　These pull chips up and away from the material. These bits leave a rougher finish at the top, but a smooth finish at the bottom of a pocket. They are ideal for plastics like HDPE and acrylic.

Downcut flute
　These pull chips down and into the material. They provide a downward force, which helps reduce part lifting. Sometimes, with a material like acrylic, you can get chip rewelding if there is no space below the part for chip expansion.

Compression flute
　This style has spirals in both directions—up and down—to eliminate chipping on both the top and bottom of the workpiece. It is typically used for laminated materials and produces a good top and bottom finish on the part.

Number of flutes

Single flute
　Allows for larger chiploads in softer materials.

Double flute
　Allows for a better part finish in harder materials.

Multiple flutes
　Allows for an even better part finish in harder materials.

To calculate your own speeds and feeds, a good rule of thumb is to use the following formula:

　Feed Rate = Spindle Speed * Chip Load * Number of Flutes

The manufacturer of the bits publishes the chip load for each bit. You count the number of flutes by looking at the bit. The spindle speed of Carvey is 12,500 RPM. The spindle speed on an

X-Carve using a Dewalt 611 can range from 16,000–27,000 RPM. I always keep the DeWalt on setting 1, which is 16,000 RPM.

The Inventables team has made available a companion video episode of Easel Live on bits at *https://www.inventables.com/projects/carving-bits-101*. For even more information, Bob Warfield created the website CNCcookbook.com. On this site, you'll find an app called G-Wizard Calculator that calculates the correct settings for each type of bit.

Early Access to 3D Profiling in Easel

As of this writing, the public version of Easel supports 2.5D designs and importing G-code for 2D, 2.5D, or 3D designs.

If you want to do a 3D design, you will need to design it in another program that includes CAD and CAM capabilities. Some examples of these programs include Fusion 360, PTC Creo, and Vectric Aspire. Alternatively, you can design in a CAD program like SolidWorks, Onshape, Rhinoceros, or CATIA and then generate toolpaths in a CAM program like MasterCAM, HSM Works, or Edgecam.

In our R&D version of Easel, we are building a 3D profiling feature. We are still working on it but hope to launch it soon. As a reader of this book, we will give you early access to this feature before we launch it publicly. When it is available for beta testing, you'll be able to go to the Machine menu and check the box to enable it. Figure 8-3 shows some example shapes you will be able to make.

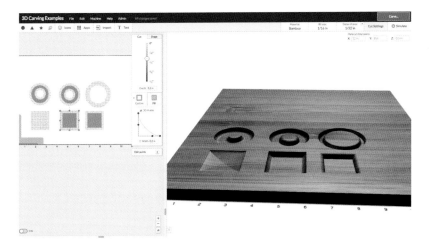

Figure 8-3. *Some of the new 3D shapes you'll be able to make in Easel*

We're always improving Easel and adding new features. We encourage you to check it out and see what's new!

You can sign up to be notified about new features at *https://www.inventables.com/newsletter_signup*.

Where Can You Go from Here?

This book is designed to get you carving as quickly as possible, but we have just scratched the surface. For me, 3D carving was a gateway into mechanical engineering. During high school, I built a scale model roller coaster out of steel that stood 5′ tall and had a 10′ x 12′ footprint. It took me two years to complete. Some students were told they should go into engineering because they were good at math. I wasn't one of those kids. I got into engineering because I liked to build things. As I built my roller coaster, I learned how to use CAD and CAM software, how to weld, how to use a lathe, and all the necessary math. The project both inspired and informed me. Chapter 9 is designed to inspire you to find your roller coaster. We'll walk through some interesting projects from the community.

9/What's Next

You've started. I hope you continue the momentum.

I believe that the world of physical product design and manufacturing is at the beginning of a major transition. It's a transition that began when the first computer was connected to a milling machine in the 1950s. An industry shift began in 2009, and history began repeating itself. Throughout history, a technology starts out being very expensive. It begins its lifecycle in the possession of companies that can afford to pay large sums of money to have the technology developed and, oftentimes, customized to their specifications. Over time, the technology becomes standardized and less expensive. Eventually, when it's inexpensive enough, hobbyists begin using it. If useful applications are found, the technology makes it into the mainstream. At that point, a version of the technology is available to consumers, hobbyists, small businesses, consultants, and major corporations.

Making a physical product once entailed a similar cost structure. Someone, often a company or an investor, risked cash to develop the product. Today, websites like Kickstarter, Etsy, Shopify, and even Instagram and Twitter provide nearly free ways to build an audience before investing any money in your product.

At Inventables, we focus on making 3D carving as easy as possible with Easel, Carvey, and X-Carve. The bottleneck is inspiration. What are you going to make? I asked some prominent makers from the community to share what they have made with 3D carving in the hopes that these projects inspire you. The following sections present what they chose to show.

Jimmy DiResta

Jimmy DiResta (Figure 9-1) is a prolific maker. He makes stuff for a living. He has made a name for himself with his inventiveness and workshop skills, creating dozens of projects for YouTube videos and television shows such as *Hammered* and *Against the Grain* on the DIY network. He has been using tools for more than 40 years. He makes it look easy but has developed his comfort level with tools through years of experience. You can see his projects and videos on his website, *http://www.jimmydiresta.com*.

Figure 9-1. *Jimmy DiResta (photo credit Luzena Adams)*

Box Joint Toolbox

This tool tote is a design I made about 10 years ago, before I had a 3D carving machine. It's been in my shop and I use it quite a bit. The reason I decided to remake it was so that I could share the design with everyone. The design is scalable: we have a larger version (Figure 9-2) for the X-Carve, and a smaller one (Figure 9-3), for which each piece can be made on the Carvey. The files and video are on the Inventables project pages. — Jimmy

Figure 9-2. *Large box joint toolbox project for X-Carve (https://www.inventables.com/projects/diresta-tool-box)*

Figure 9-3. *Tiny box joint toolbox project for Carvey (https://www.inventables.com/projects/tiny-tool-box)*

Warren Downes

Warren Downes (Figure 9-4) hosts a YouTube channel and started One Wood, driven by his passion for making things (particularly out of wood). His YouTube channel is designed to educate and inspire people while having some fun along the way. Here's where you can find Warren and his projects:

- YouTube (*https://www.youtube.com/user/onewoodtools*)
- Facebook (*https://www.facebook.com/onewoodtools*)
- Twitter (*https://twitter.com/onewoodtools*)
- Instagram (*http://instagram.com/onewood1971/*)

Figure 9-4. *Warren Downes*

Star Wars Chocolate Mold

One thing I really enjoy is working and experimenting with different materials. This fun little project combines *Star Wars*, X-Carve, wood, food-grade silicone, and chocolate. Anytime you can work *Star Wars* and chocolate into a project, it has to be a good thing, right? For this cool project, I kept the chocolate molds simple so a beginner will have no problem tackling this project (Figure 9-5)!

You can use this project as a base for so much more. Think about materials like soaps, plaster, ice cubes, pewter, epoxy, alumilite, hard foams for props, and more. In the video, I go through how to set up the project in Easel, carve the molds on your 3D carving machine, and present some tips and trick for mixing silicone. —Warren

Figure 9-5. *Star Wars–theme molds (https://www.inventables.com/projects/star-wars-silicone-chocolate-mold)*

Zach Darmanian-Harris

Zach is cofounder of Such + Such, a design studio that strives to produce engaging and timeless furniture and products through the exploration of materials and processes. Founded in 2010 by Zach and fellow DAAP Industrial Design graduate Alex Aeschbury, Such + Such works with clients nationwide in the areas of furniture, interior design, and brand identities. For more information, please visit our studio site.

The Block Clock

The Block Clock is named after the raw blocks of wood from which the clocks are carved (Figure 9-6). With our Block Clocks, we challenge the natural material that we are carving. We fabricate each Block Clock on our 3D carving machine using sustainably harvested hardwoods. Block Clocks can be mounted with two nails or stand on their own. Our clocks are driven by a battery powered quartz movement and are individually numbered.
—Zach and Alex

Figure 9-6. *The Block Clock*

Bob Clagett

It doesn't matter what the material is, or what it's for... Bob Clagett (Figure 9-7) loves making stuff. He loves showing other people how he works to inspire and empower them to make what they're passionate about. He went to art school and spent 16 years in the software industry before he jumped ship and started making stuff for a living.

Figure 9-7. *Bob Clagett*

What's Next

Four-in-a-Row Game

There were a few games that I loved as a kid, but some of them don't hold up when trying to occupy my own children. One game my kids and I both like is Four-in-a-Row (aka Connect Four), shown in Figure 9-8. It's quick to play and accessible to a wide range of age groups. The game is one of the few games that a four-year-old can play competitively against a seven-year-old. — Bob

Figure 9-8. *Four-in-a-Row game (https://www.inventables.com/projects/connect-four-game)*

Casey Shea

Casey Shea (Figure 9-9) started the first Project Make class with Dale Dougherty and Maker Media in Sebastopol, CA, in 2011. The next year, the class moved into a 3,200-ft^2 shop on the Analy High School campus, and doubled in size. In addition to teaching Project Make, Casey works with the Sonoma County Office of Education and Sonoma State University to help teachers establish and grow their making programs.

Figure 9-9. *Casey Shea*

The Mighty Morphmobile

The Mighty Morphmobile (Figure 9-10) is a reconfigurable rolling vehicle platform. Students use it to tinker around with different propulsion mechanisms, gather data, and participate in design challenges. Over the years, they have used materials that required only utility knives to fabricate, such as cardboard or balsa wood.

Figure 9-10. *The Mighty Morphmobile (https://www.inventables.com/projects/the-mighty-morphmobile)*

Steve Carmichael

Steve (Figure 9-11) hosts a YouTube channel called *The Carmichael Workshop*, where he makes fun woodworking projects on his wife's side of the garage. Steve's father introduced him to woodworking at a young age, and it has become a lifelong hobby. He is also a musician and likes to combine these two hobbies whenever he gets the chance. You can visit Steve at The Carmichael Workshop (*http://www.thecarmichaelworkshop.com/*) and follow him on social media to keep up with what's happening in his shop!

- YouTube (*https://www.youtube.com/user/carmichaelworkshop*)
- Facebook (*https://www.facebook.com/thecarmichaelworkshop*)
- Twitter (*https://www.twitter.com/carmichaelwkshp*)
- Instagram (*http://www.instagram.com/carmichaelwkshp/*)

Figure 9-11. *Steve Carmichael*

Making an Electric Guitar

In this project, you will carve a guitar body, neck, and fret board from wood blanks using the files provided. Then, you assemble these parts (along with a few other ready-to-use guitar parts and electronic components) to complete the guitar (Figure 9-12).

Figure 9-12. *The finished guitar (https://www.inventables.com/projects/electric-guitar)*

It's a Special Time in the World

Before the desktop publishing revolution in 1977, there were about 100,000 books published. Authors worked with publishing companies like Random House and Hyperion to publish their books. In a few short years, the personal computer came out, along with desktop publishing software and printers. These new tools made it possible for anyone with a computer to type a book. In 2010, about 300,000 books were published through traditional publishers and an additional 2,700,0000 books were published through on-demand publishers. Accessibility to publishing tools changed the publishing industry forever.

The development and growth of 3D carving shares similar charateristics with desktop publishing, but the world of computers is far more advanced. At Inventables, we believe that software has the potential to level the playing field when it comes to helping people make products. In the same way desktop publishing leveled the playing field with respect to who could write books, we hope our software helps people make products. In the past, you needed a publisher to believe in your book and pay an advance.

Today, you can use a free computer at the library and write your book without getting permission from any publisher. You can publish an ebook and make it available to millions for effectively no hard costs. Similarly, in the past, you needed specialized skills and a factory to make a product. Easel, Carvey, and X-Carve make it possible to make products at home or school. You can make signs, toolboxes, or even a guitar using these accessible tools and technologies.

It's now up to you.

Welcome to the community.

Index

Symbols

2D/2.5D/3D, 99
3D printing, 93

A

Aeschbury, Alex, 112

B

bias toward action, 95
bits
    additional bit option, 28
    broken, 27
    changing, 87
    choosing diameter of, 23-26
    descriptions of, 102
    materials, 102
    sizes, 11, 53
    using your own, 12
bonding wood, 51-52
borderless carving option, 80
Brown, Gordon S., 96

C

CAD (computer-aided design) software, 101
calipers, digital, 21-23
Callahan, Finn, 63
CAM (computer-aided manufacturing) software, 102
Card-a-matic milling machine, 96
Carmichael, Steve, 116
carve depth, choosing, 26-28
Carvey
    changing bits, 87
    clamping with, 29
    introductory guide, xii

carving calculations, customizing, 55-56
chip load, 55, 103
Clagett, Bob, 33-34, 113
clamping, 28-31, 56, 84
CNC machines, 93-97
community challenges
    fidget spinners, 72
    stamp carving, 78, 89
community core values, 1
Computer Numerical Control (see CNC machines)
cutting boards (see inlay cutting boards)

D

Darmanian-Harris, Zach, 112
depth per pass, 13, 26, 55, 101
digital calipers, 21-23
DiResta, Jimmy, 107
double-sided tape, 31, 38, 56
Dougherty, Dale, 114
Downes, Warren, 110

E

Easel, 5-17
    3D preview, 14
    3D profiling feature, 104
    Additional bit menu, 12
    additional bit option, 28
    Apps button, 15
    Bit Size menu, 11, 16
    bits, using your own, 12
    borderless carving option, 80
    Canvas, 14
    Canvas Zoom, 14
    Carve button, 13

customizing carving calculations, 55-56
Cut Settings menu, 13, 87
Cut/Shape pane, 38
Easel button, 6
Edit menu, 7-10, 67
File menu, 7
G-code in, 11, 99-100
Help menu, 11
Home button, 14
Icon menu, 15
Import button, 82
Import menu, 15
Machine menu, 10-11
Material boundary, 15
Material dimensions, 14, 16
Material menu, 11
Pen tool, 15
Position tab, 66
Project name, 7
project window, 16
radio buttons, 16
Safety height, 10
Shape menu, 66, 87
Shape window, 16, 37, 55
Shapes, 14
Simulate button, 14, 84-85
Smart Clamp keep out area, 14
Smoothing slider, 82
Stamp Maker app, 83-84
Stepover, 10
Text menu, 15
Threshold slider, 82
Units switch, 14
user interface, 6
Window adjustment, 14
engraving the design, 27
epoxy, food-safe, 58
evolution of 3D carving, 93-98

F

feed rate, 13, 55
feedback, xx
feeds, 101
fidget spinners, 63-72
    app for, 71
    bearings, 65, 70
    carving, 69
    community challenge, 72
    designing, 65-68
    examples, 72
    files/videos for, 65
    increasing speed, 71
    measuring, 65-67
    sanding, 69
    shopping list, 64
first carve, 31
flutes, 102

G

G-code, 11, 99-100
gliders, 34-39
    carving and flying, 39
    double-sided tape with, 38
    shopping list, 35
    tabs with, 37-38

H

Hanoi Ceramic Mosaic Mural, 1
Help Guides, 11
Howie, Jim, 95

I

Inlay app, 46-47
inlay cutting boards, 41-61
    bonding, 51-52
    creating carve files, 47-50
    customizing carving calculations, 55-56
    cutting the wood, 50
    cutting thickness strips, 51
    end grain style, 42-43, 50-61
    epoxy for, 58
    examples, 61
    files/videos for, 44, 50
    inlay design, 46-46
    long grain style, 42
    pocket element, 48, 56
    protective finish, 49, 59
    rubber feet for, 60
    sanding, 53, 58

shape choices, 44-46
shopping lists, 43
inspiration, 3, 107-119
    Block Clocks, 112
    box joint toolbox, 108-110
    electric guitar, 117
    Four-in-a-row game, 114
    Morphmobile, 115
    Star Wars chocolate mold, 111

J

Jordan, Jeff, 95

K

Kaplan, Zach, 63-64

L

Lasseter, John, xix
Lucia, Travis, 77

M

Makerspaces, 19
mastery model system, 1
materials, 11

N

NC machines, 93, 97
Numerical Control (see NC machines)

P

Parsons, John T., 94-98
parts, obtaining, xii
Picciutto, David, 41-42
Plunge rate, 13, 101
Position tab, 66
profile stepover, 101
protective finishes, 49, 59

R

raster file, 81
risk, xix
RPM (revolutions per minute), 101

S

safety glasses, 27
safety height, 101
sanding, 53, 58, 69
Shape menu, 66
Shape window, 55
Shea, Casey, 114
Smart Clamp, 29, 80
Solin, Jeff, 19-20
speeds, 101
stamp carving, 77-90
    artwork, 81-83
    backer shape, 83
    carving sequence, 87-88
    community challenge, 78, 89
    companion video, 80
    creating second file, 86
    creating the relief, 83
    examples, 90
    finishing, 88
    greeting cards, 78
    setup, 80
    shopping list, 79
    simulating stamp, 84-85
    Stamp Maker app, 83-84
    template link, 80
    using the stamp, 89
stepover, 101
Stout, Bill, 94

T

tabs, 37-38, 87
terminology, 99-103
thickness, measuring, 21-23
3D printing, 93
tool change, 25
total depth of carve, 26
2D/2.5D/3D, 99
typographical conventions, xii

V

value calculation formula, 55
vector graphics, 81

X

X-Carve
changing bits, 87
clamping with, 30
introductory guide, xii

About the Author

Zach Kaplan is the founder and CEO of Inventables, the leader in 3D carving. A maker his entire life, he is on a mission to ignite digital manufacturing worldwide and provide everyone with ambition a way to get started. Inventables' flagship products Easel, Carvey, and X-Carve are used by a new wave of makers, carving everything from circuit boards to skateboards. Named a "modern Leonardo" by the Museum of Science and Industry and a "40 under 40" by *Crain's Chicago Business*, his dream is to create a world with two million digital manufacturers that have raving fans, not just customers. Kaplan has been featured on National Public Radio and has presented at the TED Conference.

Colophon

The cover and body font is BentonSans, the heading font is Serifa, and the code font is Bitstream's Vera Sans Mono.